A Primer of Ecology

A Primer of Ecology

Third Edition

NICHOLAS J. GOTELLI

University of Vermont

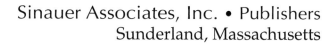

Sinauer Associates, Inc. • Publishers
Sunderland, Massachusetts

The Cover

Third-instar larvae of the North American ant lion, *Myrmeleon immaculatus*. From *Demons of the Dust* by William Morton Wheeler (W. W. Norton, New York, 1930).

The Frontispiece

A portrayal of the complexity of ecological interactions in nature. Each chapter of the primer highlights a particular interaction from this figure. Original artwork by Shahid Naeem, University of Washington.

A PRIMER OF ECOLOGY, Third Edition

Copyright © 2001 by Sinauer Associates, Inc.
All rights reserved.

This book may not be reproduced in whole or in part without permission. For information address:

Sinauer Associates, Inc., 23 Plumtree Road, Sunderland, MA 01375-0407 U.S.A.

Fax: 413-549-1118
Internet: publish@sinauer.com
www.sinauer.com

Library of Congress Cataloging-in-Publication Data
Gotelli, Nicholas J., 1959-
    A primer of ecology / Nicholas J. Gotelli. -- 3rd ed.
       p.   cm.
    Includes bibliographical references (p.).
    ISBN 0-87893-273-9 (paper)
    1. Population biology--Mathematical models. 2. Ecology--Mathematical models. I. Title
QH352.G67 2001   2001018356
577.8'8'0151--dc21   CIP

Printed in U.S.A.

6   5   4   3   2

Dedicated to my parents, Mary and Jim

Table of Contents

Chapter 3: Age-Structured Population Growth   49

Chapter 4: Metapopulation Dynamics   81

Chapter 5: Competition 99

Chapter 6: Predation 125

Chapter 7: Island Biogeography

Chapter 8: Succession

Preface to the Third Edition

In this Third Edition, I have added a new chapter (Chapter 8) that introduces students to the Markov model of ecological succession. This model provides a satisfying framework for understanding how communities change through time, and can be related to earlier verbal descriptions of facilitation, inhibition, and tolerance. Students first learned about matrix multiplication in the context of the Leslie matrix in Chapter 3; little additional effort is needed to apply this machinery to the Markov model of succession. Chapter 8 also provides more balance between population and community ecology, and a bit more coverage of plants, which some readers had requested in the first two editions.

As always, the challenge for students is to learn to think quantitatively about populations and communities. The *Primer* conveys the basic concepts and equations that underlie much of modern ecology. But this "paper-and-pencil" approach is only the first step. Words and mathematical equations are tricky, elusive things, with many nuanced levels of meaning. I think that a true understanding of these equations can only come from programming them as computer models. Computer programming requires that every detail of the model be made explicit and clear; there is no room for ambiguity or fuzziness in programming. Hilborn and Mangel's (1997) text illustrates the power of this approach for ecologists who are trying to confront the models with real data.

Learning to program, however, requires the same intellectual commitment as learning to speak a foreign language, and there is not enough time during an introductory ecology course. But help is here. Therese Donovan and Charles Welden have written a new text, called *Exercises in Ecology, Evolution, and Behavior: Programming Population Models and Simulations with Spreadsheets*. Their book teaches students how to use Excel™ spreadsheets to build all of the models in this book, as well as many other models in population genetics and evolution. It bridges the gap between ecology student and computer programmer, allowing students to "go under the hood" and see how the equations work without having to struggle with the syntax of a full-blown computer language. It is a perfect companion to this book (and an excellent tool for quantitative ecologists).

ACKNOWLEDGMENTS

Steve Jenkins and Craig Osenberg provided useful suggestions for Chapters 3 and 6, respectively, while Andy Sinauer and his staff made sure that the new material blended seamlessly with the old. I thank Gary Entsminger for extended conversations on the nature of ecological models, practical advice on computer programming, and all those guitar licks I have stolen over the years. As always, I thank my wife, Maryanne Kampmann, for her love and support, and for gardening outside my window while I wrote.

October 2000
Burlington, Vermont

Preface to the Second Edition

Authors are always tempted to add new material to a book when they get the chance to write a second edition. However, adding new models and equations to this primer would subvert its purpose, which is to concisely explain the basic models encountered in population and community ecology. In this second edition, I have added no new material, but I have included a comprehensive glossary and a short appendix on differential equations.

More importantly, I have improved the original text, corrected errors, and clarified the explanations. I am grateful to all the students and instructors who road-tested the primer and gave me their feedback. In particular, I thank Peter Bayley, Stewart Berlocher, Carol Boggs, and Sharon Strauss for sharing with me the detailed responses of their students. Tony Pakes educated me about the subtleties of stochastic growth equations, and Steve Jenkins and Juan Martinez-Gómez independently caught an insidious error in the equation for reproductive value.

Each chapter benefited from detailed reviews by a number of colleagues, including Hal Caswell, Rob Colwell, Andy Dobson, Lev Ginzburg, Bob Holt, Mark Lomolino, Bob May, Janice Moore, Mary Price, Bob Ricklefs, Joe Schall, Peter Stiling, Nick Waser, and Guiyan Yan. Chapter 5 owes a special debt to Rob Colwell. The organization of this chapter, the restatement of the competitive exclusion principle, and the "milkshake analogy" were all taken from my undergraduate lecture notes from Colwell's community ecology course at the University of California, Berkeley (winter, 1980). Shahid Naeem created original artwork for the frontispiece and chapter headings, and Neil Buckley corrected my grammar and vigilantly caught the proofing errors. Andy Sinauer and his staff transformed my text and rough drawings into a polished product. As always, I thank Maryanne for her continued love and support.

One final point. The primer has now been adopted by many colleges and universities for basic and advanced ecology courses. However, some instructors have told me that, although they enjoyed the primer, the material was too advanced for their students. Of course I am biased, but I think this attitude short-changes the students and underestimates their abilities. If the models in this primer are presented with care, they can be readily grasped by most undergraduates, even those with no previous background in ecology and little exposure to mathematics. Enrollment in ecology courses continues to climb as student awareness of environmental problems increases. Understanding the ecological principles covered in this primer is an important first step toward solving those problems.

February 1998
Burlington, Vermont

Preface to the First Edition

I love to read ecology textbooks. The latest ecology texts are well-written and entertaining to read. They cover all aspects of ecology from population growth to ecosystem ecology and conservation biology. They present students with a balanced mix of theoretical, empirical, and applied topics, supported by a vast bibliography of hundreds of literature citations—everything from the "textbook classics" to the latest cutting-edge research. All this material is packaged in an attractive format, with color photographs, sophisticated graphics, and eye-pleasing type fonts. The downside is encyclopedic length and a hefty price tag for the student.

Despite their massive size, the new texts often fail in helping students with the single most difficult aspect of ecology courses: understanding mathematical models. Many texts exclude or dilute the mathematical and quantitative material, leaving students with a product that has been intellectually gutted. More traditional texts (and instructors) that do cover mathematical models also err by assuming the mathematical details are self-evident, glossing over the derivations, and failing to explicitly and concisely state the assumptions and predictions of the models.

My own pet peeve is the treatment of the exponential model of population growth. The exponential model is the basis for most population and community models, and is often used to introduce students to concepts such as continuous versus discrete population growth, population size (N), growth rate (dN/dt), and per capita growth rate [$(1/N)(dN/dt)$]. Without a firm understanding of these ideas, students cannot grasp more complex models. Yet most textbooks devote no more than a few pages, or even a few paragraphs, to the topic of exponential population growth.

THE ORGANIZATION OF THIS BOOK

This primer grew out of my dissatisfaction with existing textbooks and the fact that I could not relegate mathematical details of the models to "course readings." In this book, I have tried to present a concise but detailed exposition of the most common mathematical models in population and community ecology. Each chapter follows the same structured format:

Model Presentation and Predictions derives the models from first principles so students can see where the equations come from. Essential equations are highlighted, but a number of intermediate algebraic "expressions" are also presented so students can understand how we get from point A to point B. With the equations in hand, the predictions from the model are explained. I have relied heavily on graphical approaches, because they are often more enlightening than algebraic solutions of the equations. Although most of the models in this book are continuous differential equations, students do not need to integrate or differentiate equations to follow this material. Instead, I have emphasized the biological interpretation of the variables in the models and how the predictions change when the variables are altered. The material in this section of each chapter is covered in some form in nearly every introductory ecology course.

Model Assumptions lists the mathematical and biological assumptions behind the equations. This material is usually covered in most textbooks, but is often scattered or buried in the text.

Model Variations explains related models that can usually be derived by relaxing one or more of the critical assumptions. In this section, I have introduced topics that are suitable for advanced and graduate-level courses, including models of environmental and demographic stochasticity, stage-structured population growth, nonlinear predator–prey isoclines, intraguild predation, and passive sampling.

Empirical Examples includes two or three field studies that illustrate the utility of the models. The examples are restricted to field studies that actually measure parameters that are relevant to the models, although in many cases I was hard-pressed to find good examples. Often, the studies in which the models fail to predict patterns in nature are more enlightening than the apparent successes.

Problems give students the chance to work with the equations and understand their behavior by plugging in some numbers. The exercises are highly simplified "story problems," but they teach students how to apply the model concepts to empirical data and give them a better intuitive understanding of the equations. Fully explained solutions are provided for each problem set. Advanced problems that correspond to the material in "Model Variations" are marked with an asterisk.

Symbols and variable names are often a source of confusion for students. I have tried to use the symbols that are encountered in most ecology textbooks, but have made some changes for clarity and consistency. The primer includes very few literature citations, which are intended only to provide sources for the equations and examples. New terms are introduced in **boldface type** and explained more thoroughly in the Glossary. The Appendix provides a brief explanation of how (and why) differential equations are used in ecology.

THE CONTENT OF THIS BOOK

Chapters 1–4 cover models for single species, and Chapters 5–7 cover models for two or more species. In Chapter 1, the model of exponential growth is developed carefully from first principles. Advanced topics include environmental and demographic stochasticity. In Chapter 2, the logistic growth model is developed as an extension of the exponential model by incorporating density dependence in birth and death rates. Discrete growth with chaos, and random and periodic variation in carrying capacity are also described. Chapter 3 covers exponential growth for age-structured populations. Advanced topics include the derivation of the Euler equation, reproductive value, and stage-structured matrix models.

Chapter 4 reflects my own interests in metapopulation models. These models relax the unrealistic assumption of no migration of individuals and represent the simplest equations for open populations. There is a close analogy between the births and deaths of individuals in a local population and the colonization and extinction of populations in a metapopulation. There is also an important conceptual link between single-species metapopulation models and the MacArthur–Wilson model of island biogeography, which is developed in Chapter 7. Although metapopulation models are only just beginning to appear in textbooks, they are an important tool for studying population dynamics in a fragmented landscape, and may have applications in conservation biology.

Chapters 5 and 6 present the standard two-species competition and predation models, and include some more complex variations with nonlinear isoclines. Chapter 5 develops a model of intraguild predation, in which species function simultaneously as predators and competitors. Chapter 6 includes a discussion of host–parasite models and briefly addresses the problem of population cycles. Both chapters stress the use of the state-space diagram as an important graphical tool for ecological modeling. Chapter 7 presents the MacArthur–Wilson equilibrium model as one possible explanation for the species–area relationship. Habitat diversity and the passive sampling model are also offered as alternative hypotheses.

PRECEDENTS FOR THIS BOOK

This book was inspired by two earlier ecology texts. The first was *A Primer of Population Biology* by E. O. Wilson and W. H. Bossert. This remarkable book, first published in 1971, has been used by thousands of students. Its concise prose, modest size, and quantitative problems introduced a generation of students to mathematical approaches in ecology and population genetics. The second was *Theoretical Ecology*, edited by R. M. May. May's overview chapters provided a concise framework for Chapters 1, 2, 5, and 6 of this primer, which cover much of the same material in a greatly expanded form.

At the risk of overstating the obvious, this primer is not a substitute for a full-length ecology text. Because of its brevity, it completely ignores many important topics in ecology that are not amenable to treatment with simple mathematical models. I hope that its concise format and modest price will justify its use as a supplementary text. If this primer helps students to understand the development, application, and limitations of mathematical models in ecology, then I will have been successful.

SOME THOUGHTS FOR THE INSTRUCTOR

I designed this primer with two sorts of courses in mind. First, the primer can serve as a supplementary text for large, introductory undergraduate courses. The material covered in "Model Presentation and Predictions" and "Model Assumptions" assumes that students have had only a single semester of calculus, and have probably forgotten most of what they learned. In my large introductory course at the University of Vermont (> 100 students), I teach all the basic material in Chapters 1, 2, 3, 5, 6, and 7. Although I do not teach the equations from Chapter 4, I do cover basic principles of metapopulations and some empirical examples. The unstarred problem sets in all the chapters are appropriate for an introductory course.

I also use the primer in my community ecology course (< 25 students), which is taught to advanced undergraduates and beginning graduate students. In this course, I treat the introductory material as a concise review, and spend more time developing the material in "Model Variations." This advanced material assumes a minimal grasp of calculus, and an exposure to basic statistical concepts of probability, means, and variances. A knowledge of matrix algebra is helpful, but not essential, for the advanced material in Chapter 3. Both the unstarred and starred problems are appropriate at this level.

My hope is that the primer will be useful to two types of instructors. Those who prefer a quantitative approach, as I do, may use the primer as a template for lectures that build ecological models from first principles. Problem-solving is essential for such a course, and most of the problems at the end of each chapter work well as exam questions.

Other instructors may not wish to devote so much lecture time to models. For these courses, the primer may serve as a tutorial to allow students to learn the details of the models on their own. In this case, instructors might wish to place more emphasis on the model assumptions and empirical examples, and perhaps eliminate the problem sets entirely.

Ecology textbooks continue to increase in size and cost, making it difficult to justify a supplemental text. However, as good as the standard textbooks are, none of them treats the mathematical models with the care and detail they deserve. I hope *A Primer of Ecology* makes your teaching easier and helps your students to better understand ecological models. For me, this has always been the most challenging and rewarding part of teaching ecology.

February 28th, 1994
5° 33' 20" N, 87° 02' 35" W
Cocos Island, Costa Rica

To the Student

The most common question beginning ecology students ask me is, "Why do we have to use so much mathematics to study ecology?" Many students enroll in my ecology course expecting to hear about whales, global warming, and the destruction of tropical rain forests. Instead they are confronted with exponential growth, doubling times, and per capita rates of increase. The two lists of topics are not unrelated. But before we can begin to solve complex environmental problems, we have to understand the basics. Just as a mechanical engineer must learn the principles of physics to build a dam, a conservation biologist must learn the principles of ecology to save a species.

The science of ecology is the study of distribution and abundance. In other words, we are interested in predicting where organisms occur (distribution), and the sizes of their populations (abundance). Ecological studies rely on measurements of distribution and abundance in nature, so we need the tools of mathematics and statistics to summarize and interpret these measurements.

But why do we need the mathematical models? One answer is that we need models because nature is so complex. We could spend a lifetime measuring different components of distribution and abundance and still not have a very clear understanding of ecology. The mathematical models act as simplified road maps, giving us some direction and idea of exactly what things we should be trying to measure in nature.

The models also generate testable predictions. By trying to verify or refute these predictions, we will make much faster progress in understanding nature than if we try to go out and measure everything without a plan. The models highlight the distinction between the *patterns* we see in nature and the different *mechanisms* that might cause those patterns.

There are two dangers inherent in the use of mathematical models in ecology. The first danger is that we build models that are too complex. When this happens, the models may contain many variables that we can never measure in nature, and the mathematical solutions may be too complex. Consequently, the most useful ecological models are often the simplest ones, and these have been emphasized throughout this primer.

The second danger is that we forget that the models are abstract representations of nature. However logical a model might appear, nothing says that nature must follow its rules. By carefully focusing on the assumptions of the model, we may be able to pinpoint the places where it departs from reality. As you will see from the examples in this primer, the models often tell us more about nature when their predictions do not match our field observations.

The purpose of this primer is to de-mystify the mathematical models used in ecology. Many of the equations in this primer can also be found in your textbook. However, your textbook may provide little or no explanation for where these equations come from, whereas the primer develops them step by step. I hope this primer will help you to understand the mathematical models and to appreciate their strengths and limitations.

CHAPTER 1

Exponential Population Growth

Model Presentation and Predictions

ELEMENTS OF POPULATION GROWTH

A **population** is a group of plants, animals, or other organisms, all of the same species, that live together and reproduce. Just as an individual grows by gaining weight, a population grows by gaining individuals. What controls population growth? In this chapter, we will build a simple mathematical model that predicts population size. In later chapters, we will flesh out this model by including resource limitation (Chapter 2), age structure (Chapter 3), and migration (Chapter 4). We will also introduce other players: populations of competitors (Chapter 5) and predators (Chapter 6) that can control growth. But for now, we will concentrate on a single population and its growth in a simple environment.

The variable N will be used to indicate the **size of the population**. Because population size changes with time, we will use the subscript t to indicate the point in time we are talking about. Thus, N_t is the number of individuals in the population at time t. By convention, we use $t = 0$ to indicate the starting point. For example, suppose we census a population of tarantulas and count 500 spiders at the beginning of our study. We revisit the population in one year and count 800 spiders. Thus, $N_0 = 500$ and $N_1 = 800$.

The units of t, in contrast to their numerical values, depend on the organism we are studying. For rapidly growing populations of bacteria or protozoa, t might conveniently be measured in minutes. For long-lived sea turtles or bristlecone pines, t would be measured in years or decades. Whatever units we use, we are interested in predicting the future population size (N_{t+1}) based on its current size (N_t).

The biological details of population growth vary tremendously among different species, and even among different populations within the same species. The factors that cause a tarantula population to increase from 500 to 800 spiders will be very different from the factors that cause an endangered condor population to decrease from 10 to 8 birds. Fortunately, all changes in population size can be classified into just four categories. Populations increase because of births and decrease because of deaths. Population size also changes if individuals move between sites. Populations increase when new individuals arrive (**immigration**) and decrease when resident individuals depart (**emigration**).

These four factors operate at different spatial scales. Births and deaths depend on current population size, as we will explain in a moment. To understand births and deaths, we need to study only the target population. By contrast, emigration and immigration depend on the movement of individuals. If we want to describe these processes, we must keep track of not just one, but several interconnected populations.

Any combination of the four factors will change population size. For our tarantula example, the initial population of 500 spiders might have produced 400 new spiderlings during the year and lost 100 adult spiders to death, with no movement of individuals. Alternatively, there might have been 50 births and 50 deaths, with 300 residents leaving (emigration) and 600 spiders arriving from other populations (immigration). Either scenario leads to an increase of 300 spiders.

These four factors can be incorporated into a mathematical expression for population growth. In this expression, B represents the number of births, D is the number of deaths, I is the number of new immigrants entering the population, and E is the number of emigrants leaving the population between time t and $t + 1$:

$$N_{t+1} = N_t + B - D + I - E \qquad \text{Expression 1.1}$$

Expression 1.1 says that population size in the next time period (N_{t+1}) equals the current population size (N_t) plus births (B) and immigrants (I), minus deaths (D) and emigrants (E). We are interested in the change in population size (ΔN), which is simply the difference in population size between last time and this time. We get this by subtracting N_t from both sides of Expression 1.1:

$$N_{t+1} - N_t = N_t - N_t + B - D + I - E \qquad \text{Expression 1.2}$$

$$\Delta N = B - D + I - E \qquad \text{Expression 1.3}$$

To simplify things, we will assume that our population is **closed**; in other words, there is no movement of individuals between population sites. This assumption is often not true in nature, but it is mathematically convenient and it allows us to focus on the details of local population growth. In Chapter 4, we will examine some models that allow for movement of individuals between patches. If the population is closed, both I and E equal zero, and we do not need to consider them further:

$$\Delta N = B - D \qquad \text{Expression 1.4}$$

We will also assume that population growth is **continuous**. This means that the time step in Expression 1.1 is infinitely small. As a consequence, population growth can be described by a smooth curve. This assumption allows us to model **population growth rate** (dN/dt) with a **continuous differential equation** (see Appendix). Thus, population growth is described as the change in population size (dN) that occurs during a very small interval of time (dt):

$$\frac{dN}{dt} = B - D \qquad \text{Expression 1.5}$$

Now we will focus on B and D. Because this is a continuous differential equation, B and D now represent respectively the **birth** and **death rates**, the number of births and deaths during a very short time interval. What factors control birth and death rates? The birth rate will certainly depend on population size. For example, a population of 1000 warblers will produce many more eggs over a short time interval than a population of only 25 birds. If each individual produces the same number of offspring during that time interval, the birth rate (B) in the population will be directly proportional to population size. Let b (lowercase!) denote the **instantaneous birth rate**. The units of b are number of births per individual per unit time [births/(individual • time)]. Because of these units, note that b is a rate that is measured **per capita**, or per individual. Over a short time interval, the number of births in the population is the product of the instantaneous birth rate and the population size:

$$B = bN \qquad \text{Expression 1.6}$$

Similarly, we can define an **instantaneous death rate** d, with units being number of deaths per individual per unit time [deaths/(individual • time)]. Of course, an individual either dies or it doesn't, but this rate is measured for a continuously growing population over a short time interval. Again, the product of the instantaneous death rate and the population size gives the population death rate:*

$$D = dN \qquad \text{Expression 1.7}$$

These simple functions will not always apply in the real world. In some cases, the birth rate may not depend on the current population size. For example, in some plant populations, seeds remain dormant in the soil for many years in a **seed bank**. Consequently, the number of emergent seedlings (births) may reflect the structure of the plant population many years ago. A model for such a population would include a **time lag** because the current growth rate actually depends on population size at a much earlier time.

Expressions 1.6 and 1.7 also imply that b and d are constant. No matter how large the population gets, individuals have the same per capita birth and death rates! But in the real world, birth and death rates may be affected by crowding: the larger the population, the lower the per capita birth rate and

---

*Note that dN in the numerator of the expression for continuous population growth (dN/dt) is *not* the same as dN in Expression 1.7. In Expression 1.7, dN is the product of the instantaneous death rate (d) and the current population size (N).

the higher the per capita death rate. We will explore this sort of **density-dependent model** in Chapter 2. For now, we will develop our model assuming a constant per capita birth rate (b) and a constant per capita death rate (d). Substituting Expressions 1.6 and 1.7 into Expression 1.5 and rearranging the terms gives us:

$$\frac{dN}{dt} = (b - d)N \qquad \text{Expression 1.8}$$

Let $b - d$ equal the constant r, the **instantaneous rate of increase**. Sometimes r is called the **intrinsic rate of increase**, or the **Malthusian parameter** after the Reverend Thomas Robert Malthus (1766–1834). In his famous "Essay on the Principle of Population" (1798), Malthus argued that food supply could never keep pace with human population growth, and that human suffering and misery were an inevitable consequence.

The value of r determines whether a population increases exponentially ($r > 0$), remains constant in size ($r = 0$), or declines to extinction ($r < 0$). The units of r are individuals per individual per unit time [individuals/(individual • time)]. Thus, r measures the per capita rate of population increase over a short time interval. That rate is simply the difference between b and d, the instantaneous birth and death rates. Because r is an instantaneous rate, we can change its units by simple division. For example, because there are 24 hours in a day, an r of 24 individuals/(individual • day) is equivalent to an r of 1 individual/(individual • hour). Substituting r back into Expression 1.8, we arrive at our first model of population growth:

$$\frac{dN}{dt} = rN \qquad \text{Equation 1.1}$$

Equation 1.1 is a simple model of **exponential population growth**. It says that the population growth rate (dN/dt) is proportional to r and that populations only increase when the instantaneous birth rate (b) exceeds the instantaneous death rate (d), so that $r > 0$. If r is positive, population growth continues unchecked and is proportional to N: the larger the population, the faster its rate of increase.

When will our model population not grow? A population will neither increase nor decrease when the population growth rate equals zero ($dN/dt = 0$). For Equation 1.1, there are only two cases when this is true. The first is when $N = 0$. Because of migration, population growth in nature will not necessarily stop when the population reaches zero. But in our simple model immigration is not allowed, so the population will stop growing if it ever hits the "floor" of zero individuals. The population will also stop growing if r should equal zero. In other words, if the per capita birth and death rates are exactly balanced, the population will neither increase nor decrease in size. In all other cases, the population will either increase exponentially ($r > 0$) or decline to extinction ($r < 0$).

PROJECTING POPULATION SIZE

Equation 1.1 is written as a differential equation. It tells us the population growth rate, but not the population size. However, if Equation 1.1 is integrated (following the rules of calculus; see Appendix), the result can be used to project, or predict, population size:

$$N_t = N_0 e^{rt} \qquad \text{Equation 1.2}$$

N_0 is the initial population size, N_t is the population size at time t, and e is a constant, the base of the natural logarithm ($e \approx 2.718$). Knowing the starting population size and the intrinsic rate of increase, we can use Equation 1.2 to forecast population size at some later time. Equation 1.2 is similar to the formula used by banks to calculate compound interest on a savings account.

Figure 1.1a illustrates some population trajectories that were calculated from Equation 1.2 for five different values of r. In Figure 1.1b, these same data are shown on a semilogarithmic plot, in which the y axis is the natural logarithm (base e) of population size. This transformation converts an exponential growth curve to a straight line. The slope of this line is r.

These graphs show that when $r > 0$, populations increase exponentially, and that the larger the value of r, the faster the rate of increase. When $r < 0$, populations decline exponentially. Mathematically, such populations never truly reach zero, but when the population reaches a projected size of less than one individual, extinction has occurred (by definition).

CALCULATING DOUBLING TIME

One important feature of a population (or a savings account) that is growing exponentially is a constant **doubling time.** In other words, no matter how large or small the population, it will always double in size after a fixed time period. We can derive an equation for this doubling time, t_{double}, by noting that, if the population has doubled in size, it is twice as large as the initial population size:

$$N_{t_{double}} = 2N_0 \qquad \text{Expression 1.9}$$

Substituting back into Equation 1.2 gives an expression in terms of initial population size:

$$2N_0 = N_0 e^{rt_{double}} \qquad \text{Expression 1.10}$$

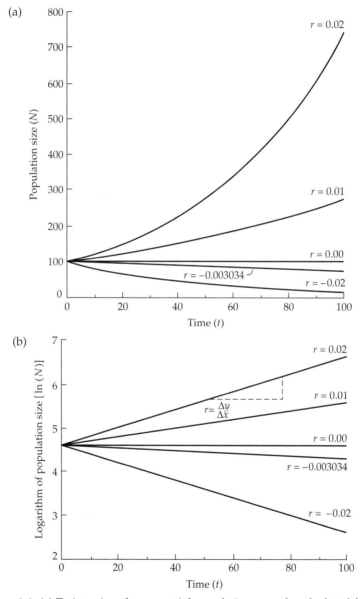

Figure 1.1 (a) Trajectories of exponential population growth, calculated from a starting population size of 100 individuals. The estimated r of -0.003034 [individuals / (individual • year)] corresponds to the projection for the grizzly bear (*Ursus arctos horribilis*) population of Yellowstone National Park (see Figure 1.6). (b) Exponential growth curves plotted on a semilogarithmic graph. The same data are used as in (a), but the y axis (population size) shows the natural logarithm (base e) of population size. In this type of graph, an exponential curve becomes a straight line; the slope of that line is r, the intrinsic rate of increase.

Table 1.1 Estimates of r and doubling times for different organisms.

Species	Common name	r [individuals / (individual • day)]	Doubling time
T phage	Virus	300.0	3.3 minutes
Escherichia coli	Bacterium	58.7	17 minutes
Paramecium caudatum	Protozoan	1.59	10.5 hours
Hydra	Hydra	0.34	2 days
Tribolium castaneum	Flour beetle	0.101	6.9 days
Rattus norvegicus	Brown rat	0.0148	46.8 days
Bos taurus	Domestic cow	0.001	1.9 years
Avicennia marina	Mangrove	0.00055	3.5 years
Nothofagus fusca	Southern beech	0.000075	25.3 years

From Fenchel (1974).

Dividing through by N_0 eliminates it from both sides of the equation:

$$2 = e^{rt_{\text{double}}} \qquad \text{Expression 1.11}$$

Taking the natural logarithm of both sides gives:

$$\ln(2) = rt_{\text{double}} \qquad \text{Expression 1.12}$$

Expression 1.12 can be rearranged to solve for doubling time:

$$t_{\text{double}} = \frac{\ln(2)}{r} \qquad \text{Equation 1.3}$$

Thus the larger r is, the shorter the doubling time. Table 1.1 gives some estimated values of r (with their corresponding doubling times) for different species of plants and animals. Among species, r varies considerably, and much of this variation is related to body size: small-bodied organisms grow faster and have larger rates of population increase than large-bodied organisms. For example, bacteria and protozoa can reproduce by asexual fission every few minutes and have high population growth rates. Larger organisms, such as primates, have delayed reproduction and long generation times, which lead to low values of r. Corresponding doubling times range from minutes for viruses to decades for beech trees.

Note, however, that even "slow-growing" populations eventually will reach astronomical sizes if they increase exponentially. Table 1.2 projects the future population size for a hypothetical herd of 50 Vermont cows [$r = 0.365$

Table 1.2 Exponential growth of a herd of 50 cattle, with $r = 0.365$ cows/(cow • year).

Year	Herd size
0	50.0
1	72.0
2	103.8
3	149.5
4	215.3
5	310.1
10	1923.7
50	4.2×10^9
100	3.6×10^{17}
150	3.0×10^{25}
200	2.5×10^{33}

Population sizes calculated from Equation 1.2.

cows/(cow • year)]. After 150 years of exponential growth, the model predicts a herd of 3×10^{25} cattle, the weight of which would exceed that of the planet earth!

Model Assumptions

What are the assumptions of Equation 1.1? In other words, what is the underlying biology of a population that is growing exponentially? This is a critical question that must be asked for any mathematical model we construct. The predictions of a mathematical model depend on its underlying assumptions. If certain assumptions are violated, or changed, the predictions of the model will also change. Other assumptions may be less critical to the predictions of the model; the model is **robust** to violations of these assumptions. We make the following assumptions for a population growing according to Equation 1.1:

✔ **No I or E.** The population is "closed;" changes in population size depend only on local births and deaths. We made this simplifying assumption in Expression 1.4, so that we could model the growth of a single population without having to keep track of organisms moving between populations. In Chapter 4, we will relax this assumption and build some simple models in which there is migration between populations.

✔ **Constant b and d.** If a population is going to grow with constant birth and death rates, an unlimited supply of space, food and other resources must be available. Otherwise, the birth rate will decrease and/or the death rate will increase as resources are depleted. Constant birth and death rates also imply that b and d do not change randomly through time. Later in this chapter, we will incorporate variable birth and death rates in the model to see how the predictions are affected.

✔ **No genetic structure.** Equation 1.1 implies that all the individuals in the population have the same birth and death rates, so there cannot be any underlying genetic variation in the population for these traits. If there is genetic variation, the genetic structure of the population must be constant through time. In this case, r represents an *average* of the instantaneous rate of increase for the different genotypes in the population.

✔ **No age or size structure.** Similarly, there are no differences in b and d among individuals due to their age or body size. Thus, we are modeling a sexless, parthenogenetic population in which individuals are immediately reproductive when they are born. A growing population of bacteria or protozoa most closely approximates this situation. In Chapter 3, we will relax this assumption and examine a model of exponential growth in which individuals have different birth and death rates as they age. If there are differences among ages, the population must have a stable age structure (see Chapter 3); in this case, r is an average calculated across the different age classes.

✔ **Continuous growth with no time lags.** Because our model is written as a simple differential equation, it assumes that individuals are being born and dying continuously, and that the rate of increase changes instantly as a function of current population size. Later in this chapter, we will relax the assumption of continuous growth and examine a model with discrete generations. In Chapter 2, we will explore models with time lags, in which population growth depends not on current population size, but on its size at some time in the past.

The most important assumption on this list is that of constant b and d, which implies unlimited resources for population growth. Only if resources are unlimited will a population continue to increase at an accelerating rate. If the other assumptions are violated, populations may still increase exponentially, although migration and time lags will complicate the picture.

But unlimited resources do not occur in nature, and we know that no real population increases without bound. So, why does the exponential growth model form the cornerstone of population biology? Although no population can increase forever without limit, all populations have the *potential* for exponential increase. Indeed, this potential for exponential increase in population size is one of the key factors that can be used to distinguish living from non-living objects. The exponential model recognizes the multiplicative nature of population growth and the positive feedback that gives populations the potential to increase at an accelerating rate.

Exponential population growth is also a key feature of Charles Darwin's (1809–1882) theory of natural selection. Darwin read Malthus' writings and recognized that the surplus of offspring resulting from exponential growth would allow natural selection to operate and bring about evolutionary change. Finally, although no population can increase forever, resources may be *temporarily* unlimited so that populations go through phases of exponential increase. Outbreaks of insect pests, invasions of "weedy" plant species, and the plight of overcrowded human populations are compelling evidence of the power of exponential population growth.

Model Variations

CONTINUOUS VERSUS DISCRETE POPULATION GROWTH

We will now explore some variations on our exponential growth model. For many organisms, time does not really behave as a continuous variable. For example, in seasonal environments, many insects and annual desert plants reproduce only once, then die; the offspring that survive form the basis for next year's population. If birth and death rates are constant (as in the exponential model), then the population will increase or decrease by the same factor each year. This population has **non-overlapping generations** and is modeled with a **discrete difference equation** rather than a continuous differential equation. Suppose the population increases (or decreases) each year by a constant proportion r_d, the **discrete growth factor**. Thus, if the population increased annually by 36%, $r_d = 0.36$. The population size next year would be:

$$N_{t+1} = N_t + r_d N_t \qquad \text{Expression 1.13}$$

Combining terms gives:

$$N_{t+1} = N_t(1 + r_d) \qquad \text{Expression 1.14}$$

Let $1 + r_d = \lambda$, the **finite rate of increase**. Then:

$$N_{t+1} = \lambda N_t \qquad \text{Expression 1.15}$$

λ is always a positive number that measures the proportional change in population size from one year to the next. Thus, λ is the ratio of the population size during the next time period to the population size for the current time period (N_{t+1}/N_t). After two years, the population size will be:

$$N_2 = \lambda N_1 = \lambda(\lambda N_0) = \lambda^2 N_0 \qquad \text{Expression 1.16}$$

Notice that the "output" of Expression 1.15 (N_{t+1}) forms the "input" (N_t) for the calculation in the next time step. The general solution to this **recursion equation** after t years is:

$$N_t = \lambda^t N_0 \qquad \text{Equation 1.4}$$

Equation 1.4 is analogous to Equation 1.2, which we used to project population size in the continuous model. What does population growth look like with the discrete model? The answer depends on the precise timing of birth and death events. Imagine that births are pulsed each spring and that deaths occur continuously throughout the year. The population growth curve will resemble a jagged saw blade, with a sharp vertical increase from births each spring, followed by a gradual decrease from deaths during the rest of the year. In spite of this decrease, the overall curve will rise exponentially, because annual births exceed annual deaths (Figure 1.2). The size of each "tooth" in the growth curve will increase year after year because the same fractional increase will add more individuals to a large population than to a small one. For example, if $\lambda = 1.2$, the population increases by 20% each year.

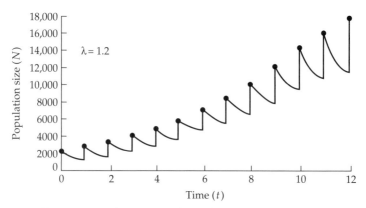

Figure 1.2 Discrete population growth. In this example, births are pulsed at the beginning of the year, and deaths occur continuously.

If the population size is 100, it will increase by 20 in one year. But when the population size is 1000, it will increase by 200 in one year.

Suppose our population reproduced twice a year, as is the case for some insects. Now we would have a "tooth" on the graph every six months. If the time step between reproductive periods becomes shorter and shorter, the teeth on the graph will be closer and closer together. Finally, if the time step is infinitely small, the curve is no longer jagged but is smooth, and we have arrived again at the continuous model of exponential growth (Equation 1.2). The continuous model essentially "connects the dots" of time in the discrete model. The continuous model is equivalent to a discrete difference equation with an infinitely small time step. Thus, we can use the rules of calculus to solve for the limit of $(1 + r_d)$ and show that:

$$e^r = \lambda \qquad \text{Equation 1.5}$$

We can express Equation 1.5 in equivalent logarithmic form as:

$$r = \ln(\lambda) \qquad \text{Equation 1.6}$$

where ln is the natural logarithm (base e). This relationship between r and λ also establishes the following numerical equivalents:

$$r > 0 \leftrightarrow \lambda > 1 \qquad \text{Expression 1.17}$$

$$r = 0 \leftrightarrow \lambda = 1 \qquad \text{Expression 1.18}$$

$$r < 0 \leftrightarrow 0 < \lambda < 1 \qquad \text{Expression 1.19}$$

Because λ is a ratio of population sizes, it is a **dimensionless number** with no units. However, λ is associated with the particular time step of the equation and cannot be changed by a simple scaling. For example, a λ of 1.2 measured with a yearly time step is *not* equivalent to a λ of 0.1 measured with a monthly time step. A λ of 1.2 yields a 20% annual increase, whereas a λ of 0.1 yields a 90% monthly decrease! If you need to change the time step for λ, first convert λ to r using Equation 1.6. Then scale r to the appropriate time units and convert back to λ with Equation 1.5. In this example, $\lambda = 1.2$ is equivalent to $r = 0.18232$ individuals/(individual • year). Dividing by 12 (months) gives $r = 0.01519$ individuals/(individual • month). From Equation 1.5, $\lambda = 1.0153$, with a monthly time step. As a check on this calculation, we can use Equation 1.4 to show that, after 12 months:

$$N_t = (1.0153)^{12} N_0 \qquad\qquad \text{Expression 1.20}$$

$$N_t = 1.2 N_0 \qquad\qquad \text{Expression 1.21}$$

This calculation demonstrates that $\lambda = 1.0153$ for a monthly time step is equivalent to $\lambda = 1.2$ for a yearly time step.

In summary, the predictions of the discrete and continuous models of exponential population growth are qualitatively similar to one another. In Chapter 2, we will see that discrete models behave very differently when we incorporate resource limitation.

ENVIRONMENTAL STOCHASTICITY

Equation 1.2 is entirely deterministic. If we know N_0, r, and t, we can calculate the predicted population size to the last decimal place. If we started over with the same set of conditions, the population would grow to precisely the same size. In such a **deterministic model**, the outcome is determined solely by the inputs, and nothing is left to chance.

Deterministic models represent an idealized view of a simple, orderly world. But the real world tends to be complex and uncertain. Think of public transportation. Does any commuter ever expect their bus or train to arrive at *precisely* the time indicated in the printed schedule? At least in American cities, buses are delayed, trains break down, and subways travel at irregular speeds, all of which introduce uncertainty (and anxiety) into the daily commute.

Could we incorporate all of the complex sources of variation into a public transportation model? Not very easily. But we could measure, each day, the arrival time of our bus. After many commuting days, we could calculate two numbers that would help us to estimate the uncertainty. The first number is the average or **mean** arrival time of the bus. If we use the variable x to indicate the time the bus arrives, the mean is depicted as $\bar{x}$. Approximately half of all buses will arrive later than $\bar{x}$ and half will arrive earlier. The second number we could calculate is the **variance** in arrival times (σ_x^2). The variance measures the variability or uncertainty associated with the mean. If the variance is small, then we know that most days the bus will arrive within, say, two minutes of the mean. But if the variance is large, the arrival time of the bus on any given morning could be as much as 20 minutes earlier *or* 20 minutes later than $\bar{x}$. Obviously, our "commuting strategy" will be affected by both the mean and the variance of x.

How can we incorporate this type of uncertainty into an exponential growth model? Suppose that the instantaneous rate of increase is no longer a simple constant, but instead changes unpredictably with time. Uncertainty in r means there are good times and bad times for population growth. During good times, the birth rate is much larger than the death rate, and the population can increase rapidly. During bad times, the difference between birth and death rates is much smaller, or perhaps even negative, so that the population increases slowly, or even decreases, for a short time period. Without specifying all of the biological causes of good and bad years, we can still develop a **stochastic** model of population growth in a varying environment. Variability associated with good and bad years for population growth is known as **environmental stochasticity**.

Imagine that a population is growing exponentially with a **mean r** ($\bar{r}$) and a **variance in r** (σ_r^2). We will use this model to predict the **mean population size** at time t ($\bar{N}_t$) and the **variance in population size** ($\sigma_{N_t}^2$). Make sure you understand the difference between these two averages and the two variances: the average and variance in r are used to predict the average and variance in N.

The derivation of this model is beyond the scope of this primer, but the results are straightforward. First, the average population size for a population growing with environmental stochasticity is:

$$\bar{N}_t = N_0 e^{\bar{r}t} \qquad \text{Equation 1.7}$$

This is no different from the deterministic model (Equation 1.2) except that we use the average r to predict the average N_t. However, like the "average family" with 2.1 children, $\bar{N}_t$ may not be a very accurate descriptor of any particular population. Figure 1.3 shows a computer simulation of a population growing with environmental stochasticity. Although the population achieves exponential increase in the long run, it fluctuates considerably from one time period to the next. The variance in population size at time t is given by (May 1974a):

$$\sigma_{N_t}^2 = N_0^2 e^{2\bar{r}t}\left(e^{\sigma_r^2 t} - 1\right) \qquad \text{Equation 1.8}$$

Other mathematical expressions for this variance are possible, depending on precisely how the model is formulated.* Equation 1.8 tells us several things about the variance of the population. First, population variance increases with time. Like stock-market projections or weather forecasts, the further

---

*Technically, we are replacing r in Equation 1.2 by $r + \sigma_r^2 W_t$, where W_t is a "white noise" distribution. This is a stochastic differential equation, which unfortunately does not have a unique solution. I have followed May (1974a), who presents the Ito solution to this problem. Biologically, the Ito solution is appropriate because it arises as a diffusion approximation to a discrete model of geometric random growth, similar to Expression 1.15. Interested readers should consult May (1973, 1974a) and Roughgarden (1979) for more details.

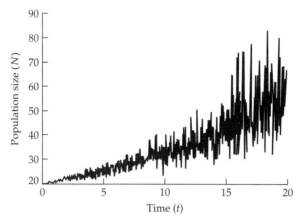

Figure 1.3 Exponential growth with environmental stochasticity. In this model, the instantaneous rate of increase fluctuates randomly through time. Here $N_0 = 20$; $r = 0.05$; $\sigma_r^2 = 0.0001$.

into the future we try to predict population size, the more uncertain our estimate. Consequently, the population growth curve resembles a funnel that flares out with increasing time (Figure 1.3). Second, the variance of N_t is proportional to both the mean and variance of r. Populations that are growing rapidly, or have a variable r, fluctuate more than slow-growing populations or those with a relatively constant r. Finally, if the variance of r is zero, Equation 1.8 collapses to zero—there is no variance in N_t, so we have returned to the deterministic model.

There is a limit to how much the population can vary in size and still persist. If N fluctuates too violently, the population may "crash" to zero. This can happen even if $\bar{r}$ is large enough to ensure rapid growth for the "average" population. Extinction from environmental stochasticity will almost certainly happen if the variance in r is greater than twice the average of r (May 1974a):

$$\sigma_r^2 > 2\bar{r} \qquad \text{Equation 1.9}$$

In our deterministic model, the population increased exponentially as long as r was greater than zero. With environmental stochasticity, the average population size also increases exponentially as a function of $\bar{r}$. However, if the variance in r is too large, there is a measurable risk of population extinction.

DEMOGRAPHIC STOCHASTICITY

Environmental stochasticity is not the only source of variability that can affect populations. Even if r is constant, populations may still fluctuate because of

demographic stochasticity. Demographic stochasticity arises, in part, because most organisms reproduce themselves as discrete units: an ostrich can lay 2 eggs or 3, but not 2.6! Some clonal plants and corals can reproduce by fragmentation and asexual budding, and in that sense, "pieces" of individuals may contribute to population increase (see Chapter 3). But for most organisms, population growth is an integer process.

If we were to follow a population over a short period of time, we would see that births and deaths are not perfectly continuous, but instead occur sequentially. Suppose that the birth rate is twice as large as the death rate. This means that a birth would be twice as likely to occur in the sequence as a death. In a perfectly deterministic world, the sequence of births and deaths would look like this: …BBDBBDBBDBBD…. But with demographic stochasticity, we might see : …BBBDDBDBBBBD…. By chance, we may hit a run of four births in a row before seeing a death in the population. This demographic stochasticity is analogous to genetic drift, in which allele frequencies change randomly in small populations.* In a model of demographic stochasticity, the probability of a birth or a death depends on the relative magnitudes of b and d:

$$P(\text{birth}) = \frac{b}{(b+d)} \qquad \text{Equation 1.10}$$

$$P(\text{death}) = \frac{d}{(b+d)} \qquad \text{Equation 1.11}$$

Suppose that, for a chimpanzee population, $b = 0.55$ births/(individual • year) and that $d = 0.50$ deaths/(individual • year). This yields an r of 0.05 individuals/(individual • year), with a corresponding doubling time of 13.86 years (Equation 1.3). From Equations 1.10 and 1.11, the probability of birth is $[0.55/(0.55 + 0.50)] = 0.524$, and the probability of death is

---

*As in the analysis of environmental stochasticity, the equations depend on the particular biological details of the model. One formulation for demographic stochasticity is that individuals in a population live and die independently of one another for random durations. Lifetimes have an exponential distribution with a mean of $1/(b + d)$. At the end of its life, an individual either replicates itself with probability $b/(b + d)$ (Equation 1.10) or it dies with probability $d/(b + d)$ (Equation 1.11). The independence of individual births and deaths leads to Equation 1.15, which gives the overall probability of population extinction.

An alternative formulation for demographic stochasticity is that change in population size is described by a matrix (Markov) transition model. In this case, the population persists with N individuals for a time that has an exponential distribution with a mean of $1/N(b + d)$. At the end of this time, the population either increases to $N + 1$ with probability $b/(b + d)$ (Equation 1.10) or it decreases to $N - 1$ with probability $d/(b + d)$ (Equation 1.11). Interested readers should consult Iosifescu and Tăutu (1973) for more details.

[0.50/(0.55 + 0.50)] = 0.476. Note that these probabilities must add to 1.0, because the only "events" that can occur in this population are births or deaths. Because a birth is more likely than a death, the chimpanzee population will, on average, increase. However, population size can no longer be projected precisely; by chance, there could be a run of births or a run of deaths in the population. Figure 1.4 shows a computer simulation of four populations that each began with 20 individuals and grew with stochastic births and deaths. Two of these populations actually declined below N_0, even though r was greater than zero.

As in our analysis of environmental stochasticity, we are interested in the average population size and its variance. The average population size at time t is again given by:

$$\overline{N}_t = N_0 e^{rt} \qquad \text{Equation 1.12}$$

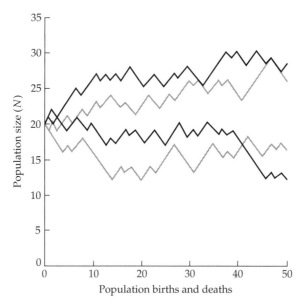

Figure 1.4 Computer simulation of population growth with demographic stochasticity. Each population track starts with an N of 20 individuals. $b = 0.55$ births / (individual • year) and $d = 0.50$ deaths / (individual • year). Although the starting conditions are identical, two of the populations actually dipped below the initial population size by the end of the simulation. Note that the x axis is not absolute time, but the number of sequential population events (births and deaths).

which is the same as in the deterministic model. The equation for variance of population size depends on whether the birth and death rates are equal or not. If b and d are exactly equal, the population will not increase on average, and the variance in population size at time t is (Pielou 1969):

$$\sigma^2_{N_t} = 2N_0bt \qquad \text{Equation 1.13}$$

If b and d are not equal, use the following:

$$\sigma^2_{N_t} = \frac{N_0(b+d)e^{rt}\left(e^{rt}-1\right)}{r} \qquad \text{Equation 1.14}$$

As in the model of environmental stochasticity, the variance in population size increases with time, and there is a risk of extinction even for populations with positive r. Demographic stochasticity is especially important at small population sizes because it doesn't take very many sequential deaths to drive a small population to extinction. Consequently, the probability of extinction depends not only on the relative sizes of b and d, but also on the initial population size. This probability of extinction is:

$$P(\text{extinction}) = \left(\frac{d}{b}\right)^{N_0} \qquad \text{Equation 1.15}$$

For the chimpanzee example, if there were 50 chimps initially, the chance of extinction would be $(0.50/0.55)^{50} = 0.009 = 0.9\%$. However, if the initial population size were only 10 chimps, the chance of extinction would be $(0.50/0.55)^{10} = 0.386 = 38.6\%$.

Equations 1.13 and 1.14 also show that demographic stochasticity depends not only on the difference between b and d, but on the absolute sizes of b and d. Populations with high birth and death rates will be more variable than populations with low rates. Thus, a population with $b = 1.45$ and $d = 1.40$ will fluctuate more than a population with $b = 0.55$ and $d = 0.50$. In both populations, $r = 0.05$, but in the first, there is a much faster turnover of individuals, and thus a much greater chance for a run of several consecutive births or deaths.

To summarize, the average population size in stochastic models of exponential growth is the same as in the deterministic model we originally derived. In a stochastic world, populations can fluctuate because of changes in the environment that affect the intrinsic rate of increase (environmental stochasticity) and because of random birth and death sequences (demographic stochasticity). For both types of variability, a population can fluctuate so much that extinction is likely, even if the average intrinsic rate of increase is positive. Demographic stochasticity is much more important as a cause of extinction at small population sizes than at large.

Empirical Examples

PHEASANTS OF PROTECTION ISLAND

Humans have introduced many species into new environments, both intentionally and accidentally. Some of these introductions have turned out to be interesting ecological experiments. For example, in 1937, eight pheasants (*Phasianus colchicus torquatus*) were introduced onto Protection Island off the coast of Washington State (Lack 1967). The island had abundant food resources and no foxes or other bird predators. The island was too far from the mainland for pheasants to fly to it, so migration did not influence population size. From 1937 to 1942, the population increased to almost 2000 birds (Figure 1.5a,b). The curve shows a jagged increase that is similar to our discrete model of population growth. This increase reflects the fact that pheasant chicks hatch in the spring, and mortality continues throughout the year.

The initial population of eight birds had increased to 30 by the beginning of 1938. If we assume that resources were not limiting growth at this time, we can estimate λ as $(30/8) = 3.75$, with a corresponding r of $\ln(3.75) = 1.3217$ pheasants/(pheasant • year). We can use this estimate to predict population size from the exponential growth model, and compare it to the actual size of the pheasant population each year. The initial predictions of this model were reasonably accurate, but after 1940, the model overestimated population size. By 1942, the population had grown to 1898 birds, whereas the model prediction was three times larger (5933 birds). This difference probably reflects depletion of food resources on the island by the increasing pheasant population. Unfortunately, this interesting ecological experiment ended abruptly when the U.S. Army set up a training camp for World War II on the island, and promptly ate the pheasants!

GRIZZLY BEARS OF YELLOWSTONE NATIONAL PARK

The grizzly bear (*Ursus arctos horribilis*) was once widespread throughout most of North America. Today, its range in the lower 48 states consists of only six fragmented populations in the northwest, some of which have fewer than 10 individuals. Yellowstone National Park supports one of the largest remaining populations, which fluctuates markedly from year to year (Figure 1.6).

The grizzly bear population data obviously do not conform to a simple exponential growth model, but they can be described by a more complex

(a)

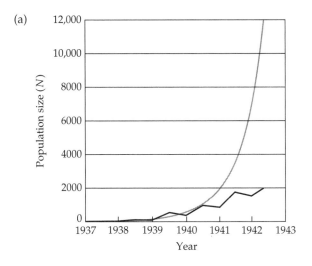

(b)

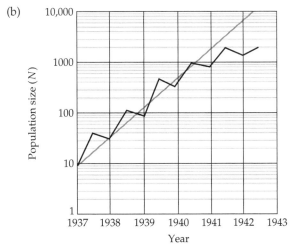

Figure 1.5 Growth of pheasant (*Phasianus colchicus torquatus*) population introduced to Protection Island. The thin line shows the hypothetical exponential growth curve, with $r = 1.3217$ individuals / (individual • year); the thick line shows the observed population size. For comparison, population sizes are plotted on a linear scale in (a) and a logarithmic scale in (b). Note that the logarithmic scale is base 10, not base e. (Data from Lack 1967.)

exponential model that incorporates environmental stochasticity (Dennis et al. 1991). The estimate of r that emerged from this model is -0.003034 bears/(bear • year), suggesting that the population will decline slowly in the long run. However, the variance for this estimate was relatively large, so we should not be surprised to see periods of population increase. Based on this

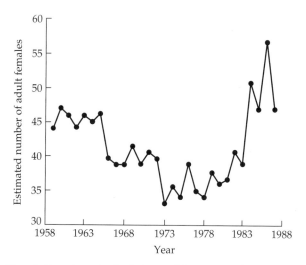

Figure 1.6 Population size of grizzly bears (*Ursus arctos horribilis*) in Yellowstone National Park. These data were used to construct a model of exponential population growth that incorporates environmental stochasticity. The estimate of *r* from this model was −0.003034 individuals / (individual • year). (From Dennis et al. 1991.)

model, the prognosis for the Yellowstone grizzly bear population is not good. The model forecasts that the population is certain to drop below 10 individuals, at which point extinction is almost guaranteed. However, because *r* is close to zero and its variance is large, the estimated time to extinction is 200 years. Thus, the model suggests that it is unlikely the grizzly bear population is in immediate danger of extinction, but that the population is likely to reach a dangerously small size in the long run.

This projection assumes that background variability in *b* and *d* will continue in the future. Thus, the model does not incorporate catastrophic events, such as the 1988 Yellowstone fire, or future changes in human activity and management strategy, such as the 1970–1971 closure of the park garbage dumps, an important food source for the bears. Because this model is one of exponential population growth in a stochastic environment, it does not incorporate resource limitation, which might lead to different predictions (see Chapter 2). Finally, the predictions of the model will change as additional data from yearly censuses become available. Increasingly, conservation biologists and park managers are using quantitative population models to estimate the risk of extinction for endangered species. Many of these models are based on the principles of exponential population growth that we have developed in this chapter.

Problems

1.1. In 1993, when the first edition of this book was written, the world's human population was expected to double in size in approximately 50 years. Assuming population growth is continuous, calculate r for the human population. If the population size in 1993 was 5.4 billion, what was the projected population size for the year 2000?

The future is here! On August 2, 2000 the best estimate of the world population size was 6.087 billion—a bit higher than that projected by the model in 1993. To find out the current estimate of the world population size, visit this website maintained by the U.S. Census Bureau:

http://www.census.gov/main/www/popclock.html

This website has a "real-time clock" that shows the estimated world and U.S. population sizes. What is today's date for you, reader, and how large is the human population now?

1.2. You are studying a population of beetles of size 3000. During a one-month period, you record 400 births and 150 deaths in this population. Estimate r and project the population size in 6 months.

1.3. For five consecutive days, you measure the size of a growing population of flatworms as 100, 158, 315, 398, and 794 individuals. Plot the logarithm (base e) of population size to estimate r.

1.4. A population of annual grasses increases in size by 12% every year. What is the approximate doubling time?

*1.5. You are studying an endangered population of orchids, for which $b = 0.0021$ births/(individual • year) and $d = 0.0020$ deaths/(individual • year). The current population size is 50 plants. A new shopping mall is planned that will eliminate part of the orchid habitat and reduce the population to 30 plants. Estimate the effect of the proposed development on the probability of extinction.

* Advanced problem

CHAPTER 2

Logistic
Population
Growth

Model Presentation and Predictions

In Chapter 1, we assumed (unrealistically) that resources for population growth were unlimited. Consequently, the per capita birth and death rates, b and d, remained constant. We did explore some models in which b and d fluctuated through time (environmental stochasticity), but those fluctuations were **density-independent**; in other words, birth and death rates did not depend on the size of the population. In this chapter, we assume that resources for growth and reproduction are limited. As a consequence, birth and death rates depend on population size. To derive this more complex **logistic growth model**, we will start with the familiar growth equation:

$$\frac{dN}{dt} = (b' - d')N \qquad \text{Expression 2.1}$$

but now we will modify b' and d' so they are density-dependent and reflect crowding.

DENSITY DEPENDENCE

In the face of increased crowding, we expect the per capita birth rate to *decrease* because less food and fewer resources are available for organisms to use for reproduction. The simplest formula for a decreasing birth rate is a straight line (see Figure 2.1):

$$b' = b - aN \qquad \text{Expression 2.2}$$

In this expression, N is population size, b' is the per capita birth rate, and b and a are constants. From Expression 2.2, the larger N is, the lower the birth rate. On the other hand, if N is close to zero, the birth rate is close to b. The constant b is the birth rate that would be achieved under ideal (uncrowded) conditions, whereas b' is the actual birth rate, which is reduced by crowding. Thus, b has the same interpretation as in the original exponential growth model: it is the instantaneous per capita birth rate when resources are unlimited. The constant a measures the strength of density dependence. The larger a is, the more sharply the birth rate drops with each individual added to the population. If there is no density dependence, then $a = 0$, and the birth rate equals b, regardless of population size. Thus, the exponential growth model is a special case of the logistic model in which there are no crowding effects on the birth rate ($a = 0$) or on the death rate ($c = 0$).

Using similar reasoning, we can modify the death rate to reflect density dependence. In this case, we expect the death rate to *increase* as the population grows:

$$d' = d + cN \qquad \text{Expression 2.3}$$

Again, the constant d is the death rate when the population size is close to zero, and the population is growing (almost) exponentially. The constant c measures the increase in the death rate from density dependence.

Expressions 2.2 and 2.3 are the simplest mathematical descriptions of the effects of crowding on birth and death rates. In real populations, the functions may be more complex. For example, b' and d' may not decline in a linear fashion; instead, there may be no change in b' or d' until a critical threshold density is reached. Some animals can reproduce, hunt, care for their offspring, or avoid predators more efficiently in groups than they can by themselves. For these populations, b' may actually increase and d' decrease as the population grows. This **Allee effect** (Allee et al. 1949) is usually important when the population is small, and may generate a critical minimum population size, below which extinction occurs (see Problem 2.3). But as the population grows, we expect negative density effects to appear as resources are depleted.

Note that *both* birth and death rates are density-dependent in this model. But it might be that only the death rate is affected by population size, and the birth rate remains density-independent, or vice versa. Fortunately, the algebra of this case works out exactly the same (see Problem 2.5). As long as either the birth rate *or* the death rate shows a density-dependent effect, we arrive at the logistic model.

Now we substitute Expressions 2.2 and 2.3 back into 2.1:

$$\frac{dN}{dt} = [(b - aN) - (d + cN)]N \qquad \text{Expression 2.4}$$

After rearranging the terms:

$$\frac{dN}{dt} = [(b - d) - (a + c)N]N \qquad \text{Expression 2.5}$$

Next, we multiply Expression 2.5 by $[(b - d)/(b - d)]$. This term equals 1.0, so it does not change the results, but allows us to simplify further:

$$\frac{dN}{dt} = \left[\frac{(b-d)}{(b-d)}\right][(b - d) - (a + c)N]N \qquad \text{Expression 2.6}$$

$$\frac{dN}{dt} = [(b - d)]\left[\frac{(b-d)}{(b-d)} - \frac{(a+c)}{(b-d)}N\right]N \qquad \text{Expression 2.7}$$

Treating $(b - d)$ as r, we have:

$$\frac{dN}{dt} = rN\left[1 - \frac{(a+c)}{(b-d)}N\right] \qquad \text{Expression 2.8}$$

CARRYING CAPACITY

Because a, c, b, and d are all constants in Expression 2.8, we can define a new constant K:

$$K = \frac{(b-d)}{(a+c)}$$

Expression 2.9

The constant K is used for more than just mathematical convenience. It has a ready biological interpretation as the **carrying capacity** of the environment. K represents the maximum population size that can be supported; it encompasses many potentially limiting resources, including the availability of space, food, and shelter. In our model, these resources are depleted incrementally as crowding increases. Because K represents maximum sustainable population size, its units are numbers of individuals. Substituting K back into Expression 2.8 gives:

$$\frac{dN}{dt} = rN\left(1 - \frac{N}{K}\right)$$

Equation 2.1

Equation 2.1 is the logistic growth equation, which was introduced to ecology in 1838 by P.-F. Verhulst (1804–1849). It is the simplest equation describing population growth in a resource-limited environment, and it forms the basis for many models in ecology.

The logistic growth equation looks like the equation for exponential growth (rN) multiplied by an additional term in parentheses ($1 - N/K$). The term in parentheses represents the **unused portion of the carrying capacity**. As an analogy, think of the carrying capacity as a square frame that will hold a limited number of flat tiles, which are the individuals. If the population should ever exceed the carrying capacity, there would be more tiles than could fit in the frame. The unused portion of the carrying capacity is the percentage of the area of the frame that is empty (Krebs 1985).

For example, suppose $K = 100$ and $N = 7$. The unused portion of the carrying capacity is $[1 - (7/100)] = 0.93$. The population is relatively uncrowded and is growing at 93% of the growth rate of an exponentially increasing population [$rN(0.93)$]. In contrast, if the population is close to K ($N = 98$), the unused carrying capacity of the environment is small: $[1 - (98/100)] = 0.02$. Consequently, the population grows very slowly, at 2% of the exponential growth rate [$rN(0.02)$]. Finally, if the population should ever exceed carrying capacity ($N > K$), the term in parentheses becomes negative, which means that the growth rate is less than zero, and the population declines towards K. Thus, density-dependent birth and death rates provide an effective brake on exponential population growth.

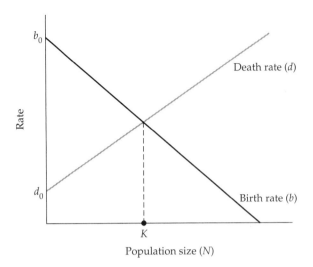

Figure 2.1 Density-dependent birth and death rates in the logistic model. The graph illustrates how the per capita rates of birth and death change as a function of crowding. The population reaches a stable equilibrium ($N = K$) at the intersection of the curves, where birth and death rates are equal.

When will the population stop growing? As in the exponential model, the rate of population growth (dN/dt) is zero when either r or N equals zero. But in the logistic model, the population will also stop growing when $N = K$. This is illustrated in Figure 2.1, which shows the density-dependent birth and death functions in the same graph. The two curves intersect at the point $N = K$ and form a **stable equilibrium**. The equilibrium is stable because no matter what the starting size of the population, it will move towards K. If N is less than K, we are at a point to the left of the intersection of the birth and death curves. In this region of the graph, the birth rate exceeds the death rate, so the population will increase. If we are to the right of the intersection point, the death rate is higher than the birth rate, and the population will decline (see Appendix).

As with the exponential growth model, we can use the rules of calculus to integrate the growth equation and express population size as a function of time:

$$N_t = \frac{K}{1 + \left[(K - N_0)/N_0\right]e^{-rt}} \qquad \text{Equation 2.2}$$

From Equation 2.2, the graph of N versus time for logistic growth is a characteristic S-shaped curve (Figure 2.2). When the population is small, it

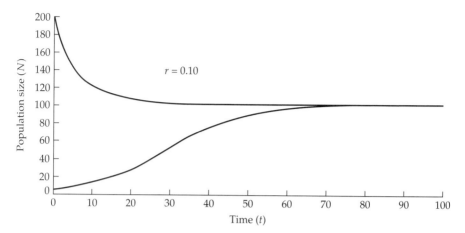

Figure 2.2 Logistic growth curve. The graph of N versus time increases in a characteristic S-shaped fashion when the population begins below carrying capacity. Above carrying capacity, the curve drops rapidly to the equilibrium point. In this example, $K = 100$, and the starting population size is 5 or 200.

increases rapidly, at a rate slightly less than that predicted by the exponential model. The population grows at its highest rate when $N = K/2$ (the steepest point on the curve), and then growth decreases as the population approaches K (Figure 2.3a). This is in contrast to the exponential model, in which the population growth rate increases linearly with population size (Figure 2.3b). In the logistic model, if the population should begin above K, Equation 2.1 takes on a negative value, and N will decline towards carrying capacity.

Regardless of the initial number of individuals (N_0), a population growing according to the logistic model will quickly reach a fixed carrying capacity, which is determined solely by K. However, the time it takes to reach that equilibrium is proportional to r; faster-growing populations reach K more quickly.

Model Assumptions

Because the logistic model is derived from the exponential model, it shares the assumptions of no time lags, migration, genetic variation, or age structure in the population. But resources are limited in the logistic model, so we make two additional assumptions:

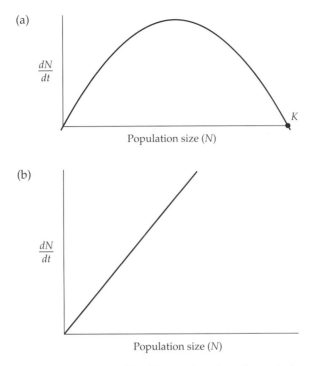

Figure 2.3 Population growth rate (dN/dt) as a function of population size. (a) Logistic growth. (b) Exponential growth.

✔ *Constant carrying capacity.* In order to achieve the S-shaped logistic growth curve, we must assume that K is a constant: resource availability does not vary through time. Later in this chapter, we will relax this assumption.

✔ *Linear density dependence.* The logistic model assumes that each individual added to the population causes an incremental decrease in the per capita rate of population growth. This is illustrated in Figure 2.4a, which shows the **per capita population growth rate** $(1/N)(dN/dt)$ as a function of population size. This per capita rate is at its maximum value of $(b - d) = r$ when N is close to zero, then declines linearly to zero when N reaches K. If N exceeds K, the per capita growth rate becomes negative. Although b and d are constants, the actual birth and death rates (b' and d') now change as a function of population size (Expressions 2.2 and 2.3). In contrast, the corresponding graph for the exponential growth model is a horizontal line because the per capita growth rate is independent of population size (Figure 2.4b).

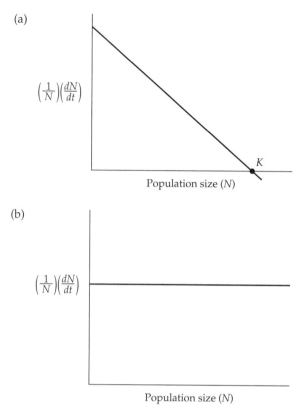

Figure 2.4 Per capita growth rates $(1/N)(dN/dt)$ as a function of population size. (a) Logistic growth. (b) Exponential growth.

Model Variations

TIME LAGS

The logistic growth model assumes that when another individual is added to the population, the per capita growth rate decreases immediately. But in many populations there may be **time lags** in the density-dependent response. For example, if a population of gulls increases in size in the fall, density dependence may not be expressed until the following spring, when females lay eggs. In a tropical rain forest, density-dependent mortality of mahogany trees (*Swietenia mahogani*) may occur in the seedling stage, but density-dependent reproduction may not occur until 50 years later, when the trees first begin to flower. Individuals do not immediately adjust their growth and reproduction when resources change, and these delays can affect population dynamics. Seasonal availability of resources, growth responses of prey populations, and age and size structure of consumer populations can introduce important time lags in population growth.

How can time lags be incorporated into our model? Suppose there is a time lag of length τ between the change in population size and its effect on population growth rate. Consequently, the growth rate of the population at time t (dN/dt) is controlled by its size at time $t - \tau$ in the past ($N_{t-\tau}$). Incorporating this time lag into the logistic growth equation gives:

$$\frac{dN}{dt} = rN\left(1 - \frac{N_{t-\tau}}{K}\right) \qquad \text{Equation 2.3}$$

The behavior of this **delay differential equation** depends on two factors: (1) the length of the time lag τ, and (2) the "response time" of the population, which is inversely proportional to r (May 1976). Populations with fast growth rates have short response times ($1/r$).

The ratio of the time lag τ to the response time ($1/r$), or $r\tau$, controls population growth. If $r\tau$ is "small" ($0 < r\tau < 0.368$), the population increases smoothly to carrying capacity (Figure 2.5a). If $r\tau$ is "medium" ($0.368 < r\tau < 1.570$), the population first overshoots, then undershoots the carrying capacity; these **damped oscillations** diminish with time until K is reached (Figure 2.5b). The exact numerical values for these trajectories are not important. What is important is to understand how the behavior of the model changes as $r\tau$ is increased.

If $r\tau$ is "large" ($r\tau > 1.570$) the population enters into a **stable limit cycle**, periodically rising and falling about K, but never settling on a single equilibrium point (Figure 2.5c). The carrying capacity is the midpoint between the high and low points in the cycle. The cycle is stable because if the population is perturbed, it will return to these characteristic oscillations. When $r\tau$ is large, the time lag is so much longer than the response time that the population repeatedly overshoots and then undershoots K. The population resembles a heating system with a faulty thermostat that constantly overheats and then overcools, never achieving an equilibrium temperature.

Cyclic populations are characterized by their **amplitude** and **period** (Figure 2.5c). The amplitude is the difference between the maximum and the average population size. It is measured on the y axis of the graph of N vs. t, and its units are number of individuals. The larger the amplitude, the greater the population fluctuations. If the amplitude is too large, the population may hit the "floor" of zero and go extinct. The period is the amount of time it takes for one complete population cycle to occur. It is measured on the x axis, in units of time. The longer the period, the greater the amount of time between population peaks.

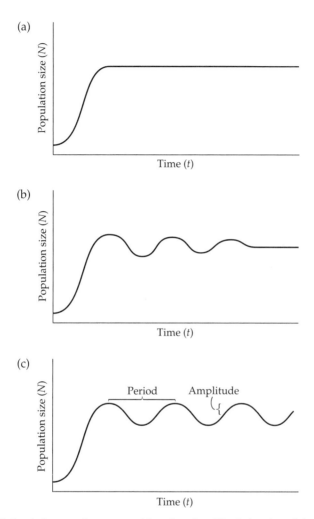

Figure 2.5 Logistic growth curves with a time lag. The behavior of the model depends on $r\tau$, the product of the intrinsic rate of increase and the time lag. (a) "Small" $r\tau$ behaves like the model with no time lag. (b) "Medium" $r\tau$ generates dampened oscillations and convergence on carrying capacity. (c) "Large" $r\tau$ generates a stable limit cycle and does not converge on the carrying capacity.

In a logistic model with a time lag, the amplitude of the cycle increases with increasing values of $r\tau$. This makes intuitive sense—if the population is growing very rapidly, or if the time lag is very long, the population will greatly overshoot K before it begins a phase of decline.

The period of the cycle is always about 4τ, regardless of the intrinsic rate of increase. Thus, a population with a time lag of one year can be expected to reach a peak density every four years. Why should the period of the cycle be four times as long as the lag? When the population reaches K, it will continue to increase for a length of time τ before starting to decrease. The distance from K to the population peak is about one-quarter of the cycle, so the length of the entire cycle is approximately 4τ. This result may explain the observation that many populations of mammals in seasonal, high-latitude environments cycle with peaks every three or four years (May 1976; see Chapter 6).

DISCRETE POPULATION GROWTH

We will now explore a model in which population growth is discrete rather than continuous. A discrete version of the logistic equation is:

$$N_{t+1} = N_t + r_d N_t \left(1 - \frac{N_t}{K}\right) \qquad \text{Equation 2.4}$$

This discrete growth logistic equation is analogous to the continuous model (Equation 2.1) in the same way that Equation 1.4 was analogous to the original exponential model (Equation 1.2). Note that the growth rate is the discrete growth factor r_d, described in Chapter 1.

A discrete population growth model has a built-in time lag of length 1.0. The population size at one time step in the future (N_{t+1}) depends on the current population size (N_t). In the last section, we saw that the product $r\tau$ controls the dynamics when a time lag is present. For the discrete model, the lag is of length 1.0, so the dynamics depend solely on r_d.

If r_d is not large, the behavior of this discrete equation is similar to that of its continuous cousin. At "small" r_d ($r_d < 2.000$), the population approaches K with damped oscillations (Figure 2.6a). At "less small" r_d ($2.000 < r_d < 2.449$), the population enters into a stable two-point limit cycle. This is similar to the continuous model, except that the population rises and falls to sharp "points," rather than following a smooth curve. The points in the discrete model correspond to peaks and valleys of the cycle (Figure 2.6b). Between an r_d of 2.449 and an r_d of 2.570, the population grows with more complex limit cycles. For example, a four-point limit cycle has two distinct peaks and two distinct valleys before it starts to repeat. The number of points in the limit cycle increases geometrically (2, 4, 8, 16, 32, 64, etc.) as the value of r_d is increased in this interval (Figure 2.6c).

But if r_d is larger than 2.570, the limit cycles break down, and the population grows in a complex, nonrepeating pattern known as **chaos** (Figure 2.6d). Mathematical models of chaos are important in many areas of science, from the description of turbulent flow to the prediction of major weather patterns.

(a)

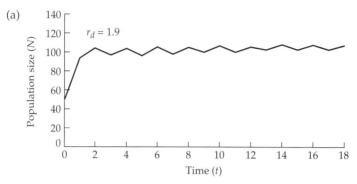

(b)

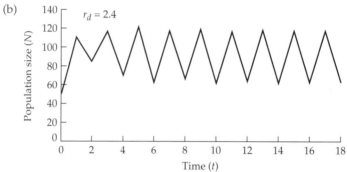

(c)

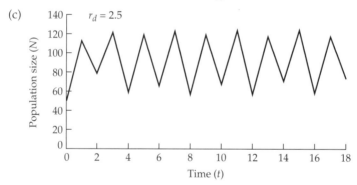

(d)

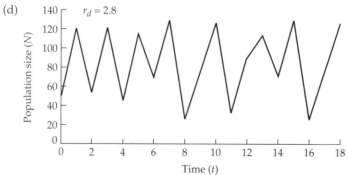

◀ **Figure 2.6** The behavior of the discrete logistic growth curve is determined by the size of r_d. (a) "Small" r_d generates damped oscillations ($r_d = 1.9$). (b) "Less small" r_d generates a stable two-point limit cycle ($r_d = 2.4$). (c) "Medium" r_d generates a more complex four-point limit cycle ($r_d = 2.5$). (d) "Large" r_d generates a chaotic pattern of fluctuations that appears random ($r_d = 2.8$).

Population biologists were among the first to appreciate that simple discrete equations may generate complex patterns (May 1974b). What is interesting about chaos is that seemingly random fluctuations in population size can emerge from a model that is entirely deterministic. Indeed, the track of a chaotic population may be so complex that it is difficult to distinguish from the track of a stochastic population.

However, chaos does not mean stochastic, or random, change. The fluctuations in a chaotic population have nothing to do with chance or randomness. Once the parameters of the model are specified (K, r_d, and N_0), the same erratic population track will be produced each time we run the model. The source of these erratic fluctuations is the density-dependent feedback of the logistic equation, combined with the built-in time lag of the discrete model. A characteristic of a chaotic population is sensitivity to initial conditions. If we alter the starting conditions, say, by changing the initial population size (N_0), the populations will diverge more and more as time goes on (Figure 2.7).

In contrast, a truly stochastic population fluctuates because one or more of its parameters (r_d or K) changes with each time step. In a stochastic model, if we alter the starting population slightly, but retain the same pattern of variation in r_d or K, the two population tracks will be slightly different, but they will not diverge as in Figure 2.7. In the next section we explore stochastic models in which the carrying capacity varies with time.

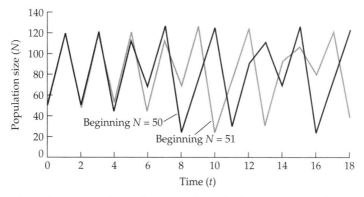

Figure 2.7 Divergence of population tracks with chaos. Both populations followed the same logistic equation, but the starting N for one of the populations was 50 and the other was 51. Note that, as more time passes, the two populations begin to diverge from one another.

RANDOM VARIATION IN CARRYING CAPACITY

In our analysis of environmental stochasticity (Chapter 1), we assumed that resources were unlimited, but that r varied randomly with time. For the logistic model, we will now assume that r is fixed, but that the carrying capacity varies randomly with time. Random variation in K means that the maximum population size that the environment can support changes unpredictably with time. How does this variation in resources affect the behavior of the logistic model? There are several mathematical approaches to the problem (May 1973; Roughgarden 1979), none of which yields a simple answer.

When r varied randomly in our exponential model, we found that the average population size was the same as in the deterministic model ($\bar{N}_t = N_0 e^{rt}$). So, you might reason that the average population size in the logistic model should approximate the average carrying capacity ($\bar{K}$). But this is not the case. Instead, $\bar{N}$ will always be *less* than $\bar{K}$. Why should this be so? When a population is above K, it declines faster than a population that is increasing from a corresponding level below K (see Problem 2.4). This asymmetry is reflected in Figure 2.2, which shows that the population tracks above and below carrying capacity are not mirror images of one another. If the carrying capacity is described by its mean ($\bar{K}$) and variance (σ_K^2) , a rough approximation to the average population size is (May 1974a):

$$\bar{N} \approx \bar{K} - \frac{\sigma_K^2}{2} \qquad \text{Equation 2.5}$$

Thus, the more variable the environment, the smaller the average population size. The pattern of population fluctuations also depends on r (Levins 1969). Populations with large r are very sensitive to changes in K, and they will tend to track these fluctuations quite closely. Consequently, the average population size will be only slightly less than the average carrying capacity. In contrast, populations with small r are relatively sluggish and will not exhibit large increases or decreases (Figure 2.8); $\bar{N}$ will be somewhat smaller than for populations with large r.

PERIODIC VARIATION IN CARRYING CAPACITY

Instead of random fluctuations in carrying capacity, suppose K varies repeatedly, in a cyclic fashion. Cyclic fluctuations in carrying capacity probably characterize many populations in seasonal temperate latitudes, and can be described with a cosine function (May 1976):

$$K_t = k_0 + k_1[\cos(2\pi t / c)] \qquad \text{Equation 2.6}$$

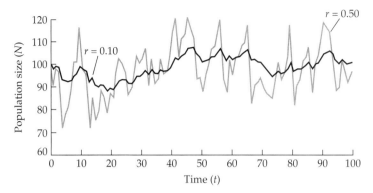

Figure 2.8 Logistic population growth with random variation in carrying capacity. Note that the population with the larger growth rate ($r = 0.50$) tracks the fluctuations in carrying capacity, whereas the population with the small growth rate ($r = 0.10$) is less variable and does not respond as quickly to fluctuations in resources.

Here, K_t is the carrying capacity at time t, k_0 is the mean carrying capacity, k_1 is the amplitude of the cycle, and c is the length of the cycle. As t increases, the cosine term in parentheses varies cyclically from −1 to 1. Thus, during a single cycle of length c, the carrying capacity of the environment varies from a minimum of $k_0 - k_1$ to a maximum of $k_0 + k_1$.

How does this cyclic variation in carrying capacity affect population growth? The length of the carrying capacity cycle functions as a kind of time lag, so once again, the behavior of the model depends on rc. If rc is small ($\ll 1.0$), the population tends to "average" the fluctuations in the environment and persists at roughly:

$$\overline{N} \approx \sqrt{k_0^2 - k_1^2} \qquad \text{Equation 2.7}$$

Thus, if rc is small, $\overline{N}$ is less than $\overline{K}$, and the reduction is greater when the amplitude of the cycle is large; both patterns are similar to the results for a population in which K varies stochastically. If rc is large ($\gg 1.0$), the population tends to track the fluctuations in the environment:

$$N_t \approx k_0 + k_1 \cos(2\pi t / c) \qquad \text{Equation 2.8}$$

although at a value slightly less than the actual carrying capacity (Figure 2.9). In conclusion, both stochastic and periodic variation in carrying capacity

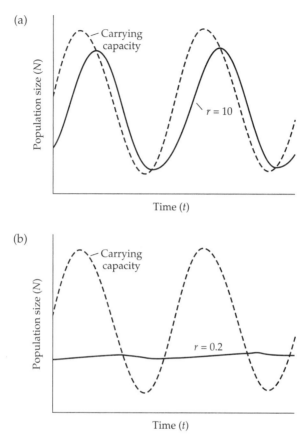

Figure 2.9 Logistic growth with periodic variation in the carrying capacity. The carrying capacity of the environment varies according to a cosine function. As with random variation, the population with the large growth rate ($r = 10$) tends to track the variation (a), and the population with the small growth rate ($r = 0.2$) tends to average it (b). The dashed line indicates K. (From May 1976.)

reduce populations, and the more variable the environment, the lower the average population size. In a variable environment, populations with large r, such as most insects, may be expected to track variation in carrying capacity, whereas populations with small r, such as large mammals, may be expected to average the environmental variation and remain relatively constant.

Empirical Examples

SONG SPARROWS OF MANDARTE ISLAND

Mandarte Island is a rocky, 6-hectare island off the coast of British Columbia. The island is home to a population of song sparrows (*Melospiza melodia*) that has been studied for many decades (Smith et al. 1991). On average, only one new female migrant joins this population each year, so most of the changes in population size are due to local births and deaths. Over the past 30 years, the population has varied between 4 and 72 breeding females and between 9 and 100 breeding males. The sparrow population of Mandarte Island does not conform to a simple logistic growth model; population size is variable and there have been periods of increase followed by rapid declines (Figure 2.10). Some of these, such as the crash in 1988, were caused by an unusually cold winter and an increased death rate. Other declines were not correlated with any obvious change in the environment.

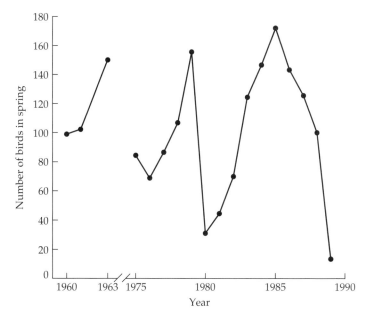

Figure 2.10 Population size of the song sparrow (*Melospiza melodia*) on Mandarte Island. (After Smith et al. 1991.)

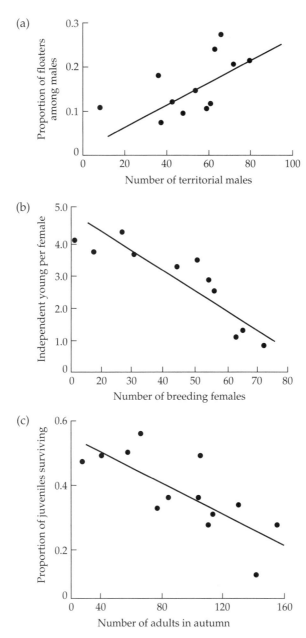

Figure 2.11 Density dependence in the Mandarte Island song sparrow (*Melospiza melodia*) population. As the population becomes more crowded (a) the proportion of nonterritorial "floater" males increases; (b) the number of surviving young produced per female decreases; (c) juvenile survival decreases. (After Arcese and Smith 1988 and Smith et al. 1991.)

Although this population is clearly buffeted by density-independent changes, there is good evidence of underlying density dependence. Male song sparrows defend territories that determine their breeding success, but limited food resources and space prevent many males from ever establishing territories. These nonterritorial "floaters" are behaviorally submissive individuals. Their proportion increased in a density-dependent fashion as the population became more crowded (Figure 2.11a). When the resident territory holders were experimentally removed, floater males quickly took over their territories, so the total breeding population size remained relatively constant.

Density dependence is also seen in the number of surviving young produced per female (Figure 2.11b), and in the survival of juveniles (Figure 2.11c), both of which decreased as the population size increased. Experimental studies confirmed that food limitation was the controlling factor: when food levels for sparrows were artificially enhanced, female reproductive output increased fourfold (Arcese and Smith 1988). Thus, both territoriality and food limitation generated density-dependent birth and death rates in song sparrows.

Nevertheless, although density dependence has the potential to control population sizes, the risk of extinction for Mandarte Island sparrows probably comes from unpredictable environmental catastrophes and other density-independent forces. Somewhat paradoxically, it is these density-independent fluctuations that allow us to detect density dependence, because they push the population above or below its equilibrium and reveal the underlying dynamics of birth and death rates.

POPULATION DYNAMICS OF SUBTIDAL ASCIDIANS

Ascidians, or "sea squirts," are filter-feeding invertebrates that live attached to pier pilings and rock walls. These animals are important components of subtidal "fouling" communities throughout the world. Ascidians are actually primitive chordates that disperse with a sexually produced tadpole larva. The perennial ascidian *Ascidia mentula* has been the subject of a long-term study of population dynamics on vertical rock walls off the Swedish west coast (Svane 1984).

Six populations were monitored continually for 12 years with photographs of permanent plots. At sheltered sites within a fjord, density was highest in shallow plots; at exposed stations, density was highest in deep-water plots. At all sites, populations fluctuated considerably (Figure 2.12), in contrast to the predictions of the basic logistic model. Mortality was primarily due to "bulldozing" by sea urchins and temperature fluctuations. These factors seemed to operate in a density-independent fashion, because there was no relationship between mortality rate and population size (Figure 2.13a). In

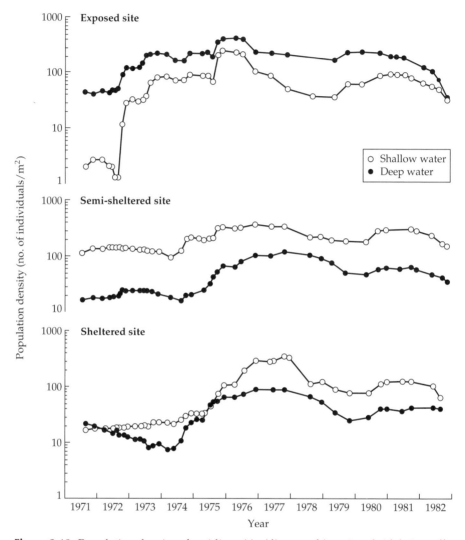

Figure 2.12 Population density of ascidians (*Ascidia mentula*) at six subtidal sites off the coast of Sweden. Population densities are greater in shallow water than in deep, except at the exposed site. Note the use of a logarithmic scale for the *y* axis, which diminishes the appearance of population fluctuations. (After Svane 1984.)

contrast, reproduction (as measured by larval recruitment) was density-dependent and decreased at high densities. At low densities, there was evidence of an Allee effect: recruitment actually increased with population density until a density of approximately 100 animals per square meter was

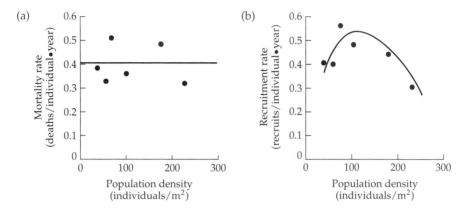

Figure 2.13 (a) Density-independent mortality rates. The mortality rate of ascidians (*Ascidia mentula*) at the six population sites appears to be independent of population size. (b) Density-dependent recruitment rates. The rate of recruitment of new juveniles into ascidian populations is density-dependent and is lower in more dense populations. Note the appearance of a possible Allee effect, as recruitment is also decreased at sites with very low abundance. (After Svane 1984.)

reached (Figure 2.13b). Possible explanations for this Allee effect include the behavioral attraction of larvae to established adults and entrapment of larvae by local water currents.

Like the Mandarte Island sparrows, these ascidians showed some evidence of underlying density dependence, although the population never reached a steady carrying capacity. Both the ascidian and sparrow populations were affected by temperature fluctuations, although these effects seemed more subtle and long-term for the ascidians. Unlike the isolated sparrow population, the ascidian populations were potentially linked by larval dispersal between sites, so that a realistic model of population dynamics might be especially complex (see Chapter 4).

LOGISTIC GROWTH AND THE COLLAPSE OF FISHERIES POPULATIONS

How many tons of fish should be harvested each year to maximize long-term yield? This **optimal yield** problem has been very important to commercial fisheries because of the huge amounts of money involved and because overfishing has been a problem since at least the 1920s, when commercial stocks of many species started to decline. The logistic growth curve provides a simple, though often unpopular, prescription for optimal fishing strategies.

The optimal strategy is the one that maximizes the population growth rate, because this rate determines how quickly fish can be removed from the pop-

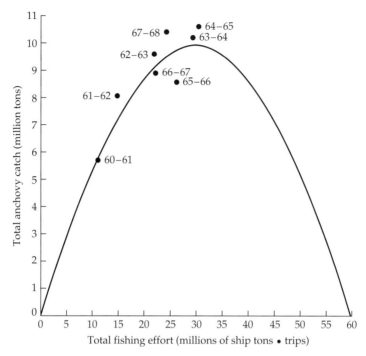

Figure 2.14 Relationship between fishing effort and total catch for the Peruvian anchovy (*Engraulis ringens*) fishery. Each point represents the fishing catch and effort for a particular year. The data include fishing effort by humans and fish catches by seabird populations. The parabola is drawn by fitting the logistic model to data from Boerema and Gulland (1973). (After Krebs 1985.)

ulation while still maintaining a constant stock for future production. If a population is growing according to the logistic equation, maximum population growth rate occurs if the population is held at $K/2$, half the carrying capacity (Figure 2.3a). Two other strategies are guaranteed to produce low yields. The first is to be extremely conservative and remove very few animals at each harvest. This keeps the standing stock large, but the yield is low because the population is close to carrying capacity and grows slowly. The other strategy is to harvest the population down to a very small size. This also produces low yield because there are so few individuals left to reproduce.

Unfortunately, this latter strategy of overdepletion has been followed by all the world's fisheries. Figure 2.14 shows the yearly catch of Peruvian anchovy (*Engraulis ringens*) fitted to the predictions of a simple logistic model. The model predicts a maximum sustained yield of approximately 10 to 11

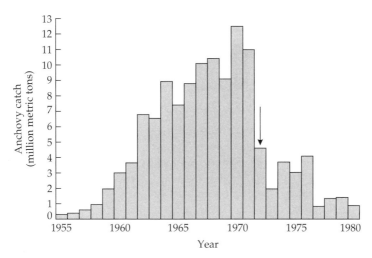

Figure 2.15 Total catch for the Peruvian anchovy (*Engraulis ringens*) fishery from 1955 to 1981. This was the largest fishery in the world until its collapse in 1972. (After Krebs 1985; unpublished data from M. H. Glanz.).

million metric tons per year. The annual catch was close to this sustained maximum from 1964 to 1971. In 1972, the Peruvian anchovy fishery collapsed, in part due to overfishing, and in part due to an El Niño event, in which a warm tropical water mass moved off the coast of Peru and greatly reduced productivity. Although fishing was reduced to allow stocks to recover, anchovy populations have never reached their former abundance and fishing yields remain low (Figure 2.15). Increasingly sophisticated technology and large factory-ships have depleted world stocks of many fish populations to the point where the industry itself is doomed to economic collapse. In 1989, for example, the cost of operating the world's 3 million fishing vessels was estimated at $92 billion, whereas the total catch was worth only $72 billion (Pitt 1993). The disappearance of human societies that depend on fishing is also inevitable.

The situation can only be remedied by worldwide restrictions on fishing and short-term reductions in catch. Unfortunately, this will not be easy because each individual fishing vessel tries to maximize its short-term yield by intensive fishing. Migratory fish populations do not obey political boundaries, making international policies difficult to enforce. The problem of short-term versus long-term profits in the exploitation of natural resources is known as "the tragedy of the commons" (Hardin 1968).

Problems

2.1. Suppose a population of butterflies is growing according to the logistic equation. If the carrying capacity is 500 butterflies and $r = 0.1$ individuals/ (individual • month), what is the maximum possible growth rate for the population?

2.2. A fisheries biologist is maximizing her fishing yield by maintaining a population of lake trout at exactly 500 individuals. Predict the initial instantaneous population growth rate if the population is stocked with an *additional* 600 fish. Assume that r for the trout is 0.005 individuals/ (individual • day).

2.3. You are studying a density-dependent turtle population that has the following relationships for the birth rate b' and the death rate d' as a function of population size (N):

$b' = 0.10 + 0.03N - 0.0005N^2$
$d' = 0.20 + 0.01N$

Plot these functions in the same graph and discuss the population dynamics of the turtle. How does this model differ from the simple logistic model with linear birth and death functions?

*2.4. Prove that the decline of a population above its carrying capacity is always faster than the corresponding increase below carrying capacity. (Hint: The starting population above carrying capacity should be represented as $K + x$.)

*2.5. In our derivation of the logistic equation, we assumed that both the birth and the death rates were density-dependent. Prove that the logistic model holds for a population in which the birth rate is density-dependent and the death rate is density-independent. Use the same approach as in Expressions 2.1 to 2.9.

*2.6. Tropical populations of many organisms experience seasonal variation in rainfall and food supply, even though temperatures are fairly constant year-round. Suppose that a water-filled tropical tree hole has a carrying capacity of 500 mosquito larvae. The water level in the hole declines gradually through the dry season, so the carrying capacity varies seasonally between 250 and 750 larvae. If the population is slow-growing, what is the long-term average population size, and what sort of temporal fluctuations in population size would you expect to see? Assume that $rc \ll 1.0$.

* Advanced problem

CHAPTER 3

Age-Structured Population Growth

Model Presentation and Predictions

EXPONENTIAL GROWTH WITH AGE STRUCTURE

In Chapter 1, we represented per capita birth and death rates as constants (*b* and *d*), which allowed us to easily calculate *r* for a population with exponential growth. The resulting model was appropriate for "simple" organisms such as single-celled bacteria or protozoa. But for most plants and animals, birth and death rates depend on the *age* of an individual.

For example, a newborn elephant cannot reproduce immediately, but must grow for a decade or more before it is reproductively mature. Death rates also vary with age. Seeds, larvae, and hatchlings usually have higher mortality rates than older age classes. Death rates also tend to be high for the very oldest individuals in a population, which may be more vulnerable to predators, parasites, and disease.

The age structure of a population has the potential to affect population growth. For example, if a population consisted only of tadpoles, it would not begin to grow until the tadpoles had metamorphosed into frogs and reached sexual maturity. In contrast, if a population of monkeys consisted only of old, postreproductive individuals, it would decline to extinction.

In this chapter, we will learn how to calculate *r* for a population in which birth and death rates depend on the age of an organism. Next, we will illustrate the short-term changes in age structure of a population that occur before it settles into a pattern of steady exponential growth. We will briefly consider the problem of life history strategies—why natural selection tends to favor certain birth and death schedules. Finally, we will develop a model of population growth for organisms with complex life histories, such as corals and perennial plants, that do not exhibit simple age structure.

Many students find the analysis of life tables to be one of the most confusing topics in ecology. Admittedly, the calculations in this chapter are tedious; we have to keep track of the birth rate, death rate, and number of individuals in each age class of the population. Be careful with your subscripts, but try not to get bogged down in notation. Keep in mind that we are still using a simple model of exponential growth in an environment with unlimited resources. In that sense, the concepts presented in this chapter are no different than those in Chapter 1.

NOTATION FOR AGES AND AGE CLASSES

To begin our analysis, we need some notation to keep track of the different ages and age classes in a population. Technically, we are modeling a population with continuous births and deaths. However, because we are classifying individuals into discrete age classes, our calculations will represent approximations to continuous growth. There is more than one way to approximate

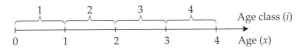

Figure 3.1 The relationship between age (x) and age class (i) in population growth models. (From Caswell 2001.)

these continuous functions, and the equations we use depend on the timing of the population censuses and the seasonal pattern of births and deaths.

We use the variable x in parentheses to refer to the **age** of an individual. For our discussion, the units of x will be years. However, any convenient time interval can be used, and the choice will usually be based on the life span of the organism and the type of census data that are available. By convention, we classify a newborn as age 0 (*not* age 1). An individual is age 0 at birth, age 0.5 at 6 months, and age 1 at its first birthday, which is the start of the second year. We use the constant k to refer to the final age in the life table, that is the age by which all individuals have died. Thus, x is a number whose value ranges from 0 to k. The number of ages in the life table depends on the length of the census interval and the life span of the organism.

Alternatively, we can designate the age of an individual by its **age class**. An individual in age class i is between the ages of $i-1$ and i (Figure 3.1). For example, an individual in the third age class is between the ages of 2 and 3. Similarly, a newborn is of age 0, but is in the first age class. If the ages in the population range from 0 to k, the age classes range from 1 to k. To keep the distinction clear, variables that indicate age will appear in parentheses, whereas variables that indicate age class will be designated by a subscript. For example, $f(5)$ indicates individuals of age 5, whereas f_5 indicates individuals in the fifth age class (those between the ages of 4 and 5).[*]

There is a subtle distinction between ages and age classes. In a continuously growing population, individuals of different ages have different birth and death rates. However, when we classify individuals into discrete age classes, we will usually be grouping individuals of slightly different ages. For example, the first age class includes both newborns and individuals who are about to celebrate their first birthday. For modeling purposes, we treat both kinds of individuals as identical and assign a single value of survival probability (P_i) and fecundity coefficient (F_i) to all individuals of an age class.

We can analyze our demographic model using the notation of either ages or age classes. We will follow the textbook tradition of using the age notation to describe the life-table analysis. However, we will switch to the age class notation to describe population growth and the analysis of complex life cycles.

---

[*]Most ecology textbooks designate ages with subscripts, but I have followed the mathematicians' convention of using subscripts for age-class matrices (see Caswell 2001).

THE FECUNDITY SCHEDULE [$b(x)$]

The **fecundity schedule** consists of the average number of female offspring born per unit time to an individual female of a particular age. The fecundity schedule is a column of values represented as $b(x)$ or $m(x)$, abbreviations for birth or maternity. For example, if $b(6) = 3$, a female of age 6 will give birth to an average of 3 female offspring. Thus, the $b(x)$ schedule gives per capita fecundity rates for females. Technically, we should be modeling the numbers of both males and females, because the two sexes often have different mortality schedules. However, we can reasonably model population growth by counting only the females.

The entries in the fecundity schedule are non-negative real numbers. An entry of zero in the fecundity schedule means that individuals of a particular age do not reproduce. The fecundity schedule gives the *average* reproduction for a female of a particular age, so these numbers do not have to be integers, and may be less than 1.0 for ages with very little reproduction.

Table 3.1 gives a hypothetical life table for an organism that lives to the end of its fourth year. The ages are 0 through 4, and the age classes are 1

Table 3.1 Standard life-table calculations.[a]

x	$S(x)$	$b(x)$	$l(x) =$ $S(x)/S(0)$	$g(x) =$ $l(x+1)/l(x)$	$l(x)b(x)$	$l(x)b(x)x$	Initial estimate $e^{-rx}l(x)b(x)$	Corrected estimate $e^{-rx}l(x)b(x)$
0	500	0	1.0	0.80	0.0	0.0	0.000	0.000
1	400	2	0.8	0.50	1.6	1.6	0.780	0.736
2	200	3	0.4	0.25	1.2	2.4	0.285	0.254
3	50	1	0.1	0.00	0.1	0.3	0.012	0.010
4	0	0	0.0		0.0	0.0	0.000	0.000
			$R_0 =$ $\Sigma l(x)b(x)$	$= 2.9$ offspring	$\Sigma = 4.3$	$\Sigma = 1.077$	$\Sigma = 1.000$	

$G = \dfrac{\Sigma l(x)b(x)x}{\Sigma l(x)b(x)}$	$= 1.483$ years
r (estimated) $= \ln(R_0)/G$	$= 0.718$ individuals/ (individual • year)
Correction added to estimated r	$= 0.058$
r (Euler)	$= 0.776$ individuals/ (individual • year)

[a] The x, $S(x)$, and $b(x)$ columns are supplied. All others are calculated from these.

through 4. We will use the data in Table 3.1 to illustrate all the calculations necessary for a typical life-table analysis. If you look at the $b(x)$ column, you see that newborns do not reproduce. One-year-olds produce an average of 2 offspring, two-year-olds produce 3 offspring, and three-year-olds produce 1 offspring.

FECUNDITY SCHEDULES IN NATURE

In nature, what sorts of fecundity schedules do we find? Animal ecologists distinguish between **semelparous** and **iteroparous** reproduction. Plant ecologists use the equivalent terms **monocarpic** and **polycarpic**. In semelparous (monocarpic), or "big bang" reproduction, an organism reproduces only once in its lifetime. Examples are oceanic salmon and many flowering desert plants. The fecundity schedule for a semelparous organism would have zeroes for all ages except for the single reproductive age. In iteroparous (polycarpic) reproduction, the individual reproduces repeatedly during its lifetime. Examples include long-lived organisms such as sea turtles and oak trees. Fecundity schedules for iteroparous organisms have non-zero entries for two or more ages.

Plant ecologists use two similar terms, **annual** and **perennial**, to refer to plants that complete their life cycle in a single season, and those that live for more than one season. Although there are many exceptions, most annual species are semelparous, and most perennial species are iteroparous. We will postpone our discussion of the evolutionary significance of these reproductive strategies. For now, we will simply use the fixed fecundity schedule for a population to help us calculate the intrinsic rate of increase.

THE SURVIVORSHIP SCHEDULE [$l(x)$]

Fecundity is only half the story. The population growth rate depends equally on the rates of mortality for different ages. Individuals of a particular age might produce dozens of offspring, but if very few individuals survive to that age, the effect on population growth rate will be minor.

How can we measure the survivorship schedule of a population? Imagine that we have a **cohort** of individuals that were all born at the same time. We follow this cohort from birth until all the individuals have died. We keep track of the number of individuals that have survived to the start of each new year. These data can be represented as a column of numbers, $S(x)$, the **cohort survival**. Table 3.1 gives some cohort data for our hypothetical life table. We begin with a cohort of 500 individuals at birth, and by the beginning of the fifth year, all of them have died.

The raw data in the $S(x)$ column must now be converted to the **survivorship schedule**, designated as $l(x)$, where l stands for life table. The quantity $l(x)$ is defined as the proportion of the original cohort that survives to the start of age x. Equivalently, we can think of $l(x)$ in terms of the survivorship of an individual. $l(x)$ is the *probability* that an individual survives from birth to the beginning of age x. To calculate $l(x)$, divide the number of survivors of age x [$S(x)$] by the size of the original cohort [$S(0)$]:

$$l(x) = \frac{S(x)}{S(0)}$$
Equation 3.1

The first entry in the $l(x)$ column is $l(0)$. It represents the survivorship of the cohort to birth. By definition, all individuals in the cohort have "survived" to the start, so the value of $l(0)$ is always 1.0 [$l(0) = S(0)/S(0) = 1.0$]. The last entry in the $l(x)$ column is $l(k)$. It represents the age that none of the original cohort reaches: $l(k)$ always equals 0.0 [$l(k) = 0.0/S(0) = 0.0$]. Between these endpoints, $l(x)$ shrinks in size as individuals in the cohort age and die. Thus, the $l(x)$ column is a set of consecutively decreasing real numbers between 1.0 and 0.0.

For the data in Table 3.1, the original cohort was 500 individuals, so we will divide each observation by this value to calculate $l(x)$. Notice that 80% of the original cohort survived to age 1 [$l(1) = 0.80$], but only 10% of the cohort made it to the start of age 3 [$l(3) = 0.10$]. This remaining 10% died between age 3 and age 4, so $l(4) = 0.0$; none of the original cohort is left.

When you calculate $l(x)$ from a survivorship schedule, take care to divide all the entries by the original cohort size [$S(0)$]. Do not make the common mistake of dividing $S(x)$ by other values in the life table. In the next section, we will calculate age-specific survival probabilities, which do use consecutive values of $S(x)$. But for the calculation of $l(x)$, always divide the observed values by $S(0)$.

SURVIVAL PROBABILITY [$g(x)$]

The survivorship schedule $l(x)$ gives the probability of survival from birth to age x. To compare the survival of different ages directly, we must determine the probability of survival from age x to age $x + 1$, *given* that an individual has already survived to age x. The **survival probability** $g(x)$ is the probability that an individual of age x survives to age $x + 1$:

$$g(x) = \frac{l(x+1)}{l(x)}$$
Equation 3.2

From Table 3.1, for example, the probability that a newborn survives its first year and reaches age 1 is $g(0) = 0.8/1.0 = 0.8$. Thus, there is an 80% chance that a newborn will still be alive at age 1. If we are thinking in terms of a cohort analysis, 80% of all newborns will be alive at age 1. In contrast, the probability of survival between ages 1 and 2 [$g(1)$] is $(0.4/0.8) = 0.5$. Although the $l(x)$ schedule never increases with age, the $g(x)$ schedule may either increase or decrease. The way in which survival probabilities change with age is an important component of the life history of an organism, as described in the next section.

SURVIVORSHIP SCHEDULES IN NATURE

What are the different types of survivorship curves observed in nature? There are three basic patterns. These can be seen by plotting the logarithm of $l(x)$ on the y axis and age (x) on the x axis. The points on this graph are connected to form a survivorship curve. The slope of this curve at any point is $\ln[g(x)]$. Therefore, if the survivorship curve forms a straight line, the probability of survival is constant over those ages.

Figure 3.2 illustrates the three types of curves. A **Type I survivorship curve** has high survivorship during young and intermediate ages, then a steep drop-off in survivorship as individuals approach the maximum life span. Examples include humans and other mammals that invest a good deal of

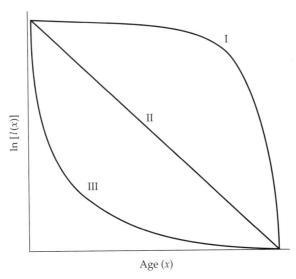

Figure 3.2 Type I, II, and III survivorship curves. Note the logarithmic transformation of the y axis.

parental care in their offspring, ensuring high survivorship of young age classes.

The opposite, and more common, pattern is a **Type III survivorship curve**. In this case, survivorship is very poor for the young age classes, but much higher for older individuals. Examples include many insects, marine invertebrates, and flowering plants. These organisms may produce hundreds or thousands of eggs, larvae, or seeds, most of which die. However, the handful of individuals that do pass through this vulnerable stage have relatively high survivorship in later years.

Finally, the **Type II survivorship curve** is intermediate between these two. Because it is a straight line on a logarithmic graph, the Type II survivorship schedule is one in which the mortality rate is constant throughout life. Few organisms have a true Type II survivorship curve, because it is unusual for the probability of death to remain constant as an organism ages. Some birds have a Type II curve for much of their lives, but often with a steeper mortality curve during the more vulnerable egg and chick stages.

The $l(x)$ and $b(x)$ schedules are the basis for all our life-table calculations. Keep in mind that these schedules are independent pieces of data about death and birth. The $l(x)$ schedule is calculated by following the survivorship of a cohort of organisms. It tells us only the chances of individuals surviving to a particular age, and contains no information about their reproduction. In contrast, the $b(x)$ schedule reveals only the per capita birth rates of females of different ages, and does not say anything about how many females actually survive to those ages. If we know the $l(x)$ and $b(x)$ schedules, we can calculate the intrinsic rate of increase, as illustrated in the next section. When you work with the $l(x)$ and $b(x)$ schedules, be careful with your notation. Remember that the $l(x)$ column gives the survivorship *up to* the start of age x, whereas the $b(x)$ schedule gives the per capita birth rates of females of age x.

CALCULATING NET REPRODUCTIVE RATE (R_0)

To estimate r from the $l(x)$ and $b(x)$ schedules, we first have to compute two other numbers, the net reproductive rate (R_0) and the generation time (G). These numbers are part of the recipe for estimating r, but they tell us important things about an age-structured population in their own right. The **net reproductive rate**, R_0, is defined as the mean number of female offspring produced per female over her lifetime. To compute R_0, multiply each value of $l(x)$ by the corresponding value of $b(x)$ and sum these products across all ages:

$$R_0 = \sum_{x=0}^{k} l(x)b(x) \qquad \text{Equation 3.3}$$

The units of R_0 are numbers of offspring. The net reproductive rate represents the reproductive potential of a female during her entire lifetime, adjusted for the mortality schedule. Suppose that there was no mortality in the population until females reached their maximum age. This would mean that $l(x) = 1.0$ for all ages except the last. In this case, Equation 3.3 would simply add up the lifetime production of offspring—the gross reproductive rate. But in most populations, mortality in each age class reduces the potential contribution of offspring to the next generation. Thus, the net reproductive rate is the offspring production discounted by mortality. For the fecundity and survivorship schedules in Table 3.1, $R_0 = 2.9$ offspring.

If R_0 is greater than 1.0, there is a net surplus of offspring produced each generation, and the population increases exponentially. If R_0 is less than 1.0, the mortality is so great that the population cannot replace itself, and it declines to extinction. Finally, if $R_0 = 1.0$, the offspring production exactly balances the mortality each generation, and the population size does not change.

This description of R_0 is very similar to the description of λ, the finite rate of increase in the exponential growth model (see Chapter 1). In fact, you might be tempted to conclude that $r = \ln(R_0)$, because $r = \ln(\lambda)$ for populations with no age structure (Equation 1.5). However, λ measures the rate of increase as a function of *absolute time*, whereas R_0 measures increase as a function of *generation time*. Therefore, if we want to calculate r, we must scale R_0 to account for generation time.

CALCULATING GENERATION TIME (G)

Generation time is a somewhat elusive concept for populations with continuous growth. Imagine that we followed a cohort from birth and kept track of all the offspring it produced. One definition of the generation time is the average age of the parents of all the offspring produced by a single cohort (Caughley 1977). This is calculated as:

$$G = \frac{\sum_{x=0}^{k} l(x)b(x)x}{\sum_{x=0}^{k} l(x)b(x)} \qquad \text{Equation 3.4}$$

The units of $l(x)$ and $b(x)$ cancel in the numerator and denominator, leaving us with an answer in units of time (x). Unless newborns have high fecundity ($b(0) >> 0$), the numerator will always be larger than the denominator in Equation 3.4. Consequently, the generation time will usually be greater than 1.0 for populations with age structure. For the data in Table 3.1, $G = 1.483$ years.

CALCULATING INTRINSIC RATE OF INCREASE (r)

We can use the equation for exponential growth to solve for r in terms of R_0 and G (Mertz 1970). Imagine a population is growing exponentially for a time G:

$$N_G = N_0 e^{rG}$$

Expression 3.1

Dividing both sides by N_0 gives:

$$\frac{N_G}{N_0} = e^{rG}$$

Expression 3.2

The ratio on the left side of the expression is an approximation to the net reproductive rate, R_0:

$$R_0 \approx e^{rG}$$

Expression 3.3

Taking the natural logarithm of both sides gives:

$$\ln(R_0) \approx rG$$

Expression 3.4

Rearranging Expression 3.4 gives us an approximation for r:

$$r \approx \frac{\ln(R_0)}{G}$$

Equation 3.5

Thus, the rate of population increase is slower for organisms with long generation times. Continuing with the data in Table 3.1, the estimate of r is 0.718 individuals/(individual • year).

Equation 3.5 is only an approximation, although it is usually within 10% of the true value (Stearns 1992). To obtain an exact solution for r, you must solve the following equation:

$$1 = \sum_{x=0}^{k} e^{-rx} l(x) b(x)$$

Equation 3.6

Equation 3.6 is adapted from the **Euler equation** (pronounced "oiler"), named after the Swiss mathematician Leonhard Euler (1707–1783), who developed it in his analyses of human demography. Later in this chapter, we will illustrate the derivation of the Euler equation. For now, we will simply use Equation 3.6 as a formula for determining the precise value of r.

Because we know the $l(x)$ and $b(x)$ schedules, the only unknown quantity in Equation 3.6 is r. Unfortunately, there is no way to solve this equation except by plugging in different values of r and adjusting your estimate upwards or downwards. A good starting place is the estimate of r from Equation 3.5. For the data in Table 3.1, substituting $r = 0.718$ into Equation 3.6 gives a sum of 1.077, whereas the correct value of r will generate a sum of exactly 1.0. This calculation indicates that our original estimate of r was too small. Because we are summing with the negative exponent of r, a larger value of r will generate a smaller sum. If we experiment with different values, we find that an r of 0.776 is a close solution to the Euler equation.

DESCRIBING POPULATION AGE STRUCTURE

Once we have calculated r from the fecundity and survivorship schedules, we can forecast the total population size by using any of the equations for exponential growth from Chapter 1. But we are also interested in knowing the number of individuals in each age class of the population. This means we will shift our notation from ages to age classes.

We will use $n_i(t)$ to indicate the number of individuals at time t in age class i. For example, if $n_1(3) = 50$, there are 50 individuals in the first age class at the third time step. Because there are k age classes in the population, the age structure at time t consists of a vector of abundances. We indicate this vector with a boldfaced, lowercase **n**:

$$\mathbf{n}(t) = \begin{pmatrix} n_1(t) \\ n_2(t) \\ \vdots \\ n_k(t) \end{pmatrix} \qquad \text{Expression 3.5}$$

For example, the vector for the population in Table 3.1 after five years might be:

$$\mathbf{n}(5) = \begin{pmatrix} 600 \\ 270 \\ 100 \\ 50 \end{pmatrix} \qquad \text{Expression 3.6}$$

Thus, there are 600 individuals in the first age class, but only 50 individuals in the terminal age class (age class 4). Using information in the mortality and fertility schedules, we can predict how the age structure of a population changes from one time period [$\mathbf{n}(t)$] to the next [$\mathbf{n}(t + 1)$].

Describing the population in terms of its age structure requires us to shift from using ages to using age classes. First, we need to obtain **survival probabilities** P_i for each age class. These probabilities represent the chance that an individual in age class i survives to age class $i + 1$. Next, we need to calculate **fertilities** F_i for each age class. These fertilities represent the average number of offspring produced by an individual in age class i. Clearly, the survivorship probabilities and fertilities for individuals of different age classes are related to the $l(x)$ and $b(x)$ schedules for individuals of different ages.

However, the conversion of these values is tricky; it depends on the timing of births and deaths within an age class, and the timing of the population census (Caswell 1989). In this primer, we will assume a simple **birth-pulse model**, in which individuals give birth to all their offspring on the day they enter a new age class. We will further assume a **postbreeding census**, in which individuals are counted each year just after they breed.

These assumptions make the calculation of P_i and F_i relatively simple. A **birth-flow model**, in which individuals reproduce continuously in an age class, would require more complex calculations. Keep in mind that the estimates of population growth will depend on how the age-class model is set up. The estimates of population growth also may not match the exact calculations from the Euler equation. Once we have the survival probabilities and fertility values for each age class, we will use them to calculate the changes in population structure with time.

CALCULATING SURVIVAL PROBABILITIES FOR AGE CLASSES (P_i)

For the birth-pulse model with a postbreeding census, the probability that an individual in age class i survives to age class $i + 1$ is:

$$P_i = \frac{l(i)}{l(i-1)} \qquad \text{Equation 3.7}$$

This equation is similar to the calculation of the age-specific survival probability $g(x)$ (Equation 3.2), although note the shift in notation as we go to a model of age classes. With Equation 3.7, it is easy to calculate the change in the number of individuals in a particular age class from one time period to the next:

$$n_{i+1}(t+1) = P_i n_i(t) \qquad \text{Equation 3.8}$$

Equation 3.8 says that the number of individuals in a particular age class next time step [$n_{i+1}(t + 1)$] is the number of individuals currently in the *previous* age class [$n_i(t)$] multiplied by the survival probability for that age class (P_i). So, the survival probability controls the rate at which individuals "graduate" to each successive age class.

CALCULATING FERTILITIES FOR AGE CLASSES (F_i)

Equation 3.8 works for all age classes except the first. The number of individuals in the first age class depends on the reproduction of all the age classes.

We define the fertility of age class i as:

$$F_i = b(i)P_i \qquad \text{Equation 3.9}$$

Equation 3.9 says that the fertility of a particular age class is the number of offspring produced, discounted by the survival probability for that age class. The discount is necessary because the parents must survive through the age class in order to reproduce and have their offspring counted.

Once F_i is known for each age class, we multiply these fertilities by the number of individuals in each age class. This product is then summed over all age classes to calculate the number of new offspring:

$$n_1(t+1) = \sum_{i=1}^{k} F_i n_i(t) \qquad \text{Equation 3.10}$$

Having derived fertility and survivorship coefficients for each age class from the $l(x)$ and $b(x)$ schedules, we can now calculate the number of individuals in each age class for a single time step. For a population with four age classes, we would have:

$$n_1(t+1) = F_1 n_1(t) + F_2 n_2(t) + F_3 n_3(t) + F_4 n_4(t)$$
$$n_2(t+1) = P_1 n_1(t)$$
$$n_3(t+1) = P_2 n_2(t)$$
$$n_4(t+1) = P_3 n_3(t)$$

Expression 3.7

In the next section we will express these changes in matrix form.

THE LESLIE MATRIX

We can represent the growth of an age-structured population in matrix form. The **Leslie matrix**, named after the population biologist Patrick H. Leslie, describes the changes in population size due to mortality and reproduction (Leslie 1945). If there are k age classes, the Leslie matrix is a $k \times k$ square matrix. It always has the following form:

$$\mathbf{A} = \begin{bmatrix} F_1 & F_2 & F_3 & F_4 \\ P_1 & 0 & 0 & 0 \\ 0 & P_2 & 0 & 0 \\ 0 & 0 & P_3 & 0 \end{bmatrix} \qquad \text{Expression 3.8}$$

Each column of the Leslie matrix is the age at time t and each row is the age at time $t + 1$. Each entry in the matrix represents a transition, or change in the number of individuals from one age class to another. In the Leslie matrix, the fertilities are always in the first row; they represent contributions to newborns from reproduction of each age class. The survival probabilities are always in the subdiagonal. They represent transitions from one age class to the next. All other entries in the Leslie matrix are 0 because no other transitions are possible. Individuals cannot remain in the same age class from one year to the next, so the diagonals must equal zero. Similarly, individuals cannot skip or repeat age classes, so other entries in the matrix are zero.

The reason for using the matrix format is that we can now describe population growth as a simple matrix multiplication:

$$\mathbf{n}(t+1) = \mathbf{An}(t)$$

Equation 3.11

In other words, the population vector in the next time step $[\mathbf{n}(t + 1)]$ equals the Leslie matrix (**A**) multiplied by the current population vector $[\mathbf{n}(t)]$. The rules of matrix algebra are used to calculate the changes in abundance in each age class, and these are equivalent to the calculations in Expression 3.7. If you have had matrix algebra, λ is the dominant eigenvalue of the Leslie matrix. Now that we have converted our age-based life-table data to an age-class Leslie matrix, we are ready to see how age structure changes during population growth.

Table 3.2 Calculation of age-specific survival probabilities and fertilities for the Leslie matrix. Data from Table 3.1. Notice that the first row of the table is blank for P_i and F_i, because we begin counting age classes at 1, not 0.

x	i	l(x)	b(x)	$P_i =$ $l(i)/l(i-1)$	$F_i =$ $b(i)P_i$
0		1.0	0		
1	1	0.8	2	0.80	1.60
2	2	0.4	3	0.50	1.50
3	3	0.1	1	0.25	0.25
4	4	0	0	0.00	0.00

The resulting Leslie matrix is:

$$\mathbf{A} = \begin{bmatrix} 1.6 & 1.5 & 0.25 & 0 \\ 0.8 & 0 & 0 & 0 \\ 0 & 0.5 & 0 & 0 \\ 0 & 0 & 0.25 & 0 \end{bmatrix}$$

STABLE AND STATIONARY AGE DISTRIBUTIONS

Table 3.2 converts the life-table data of Table 3.1 to a Leslie matrix. We use this Leslie matrix to compare the growth of two hypothetical populations. One population has 50 individuals in each age class, and the second population has 200 newborns, but no other age classes present. Figure 3.3 shows the

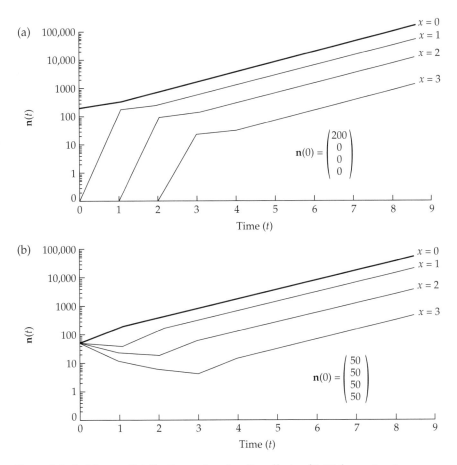

Figure 3.3 Stable age distributions, showing the effects of initial age structure on population growth. Each line represents a different age class, growing according to the birth and death schedules of Table 3.1. In (a), the initial age distribution was 200 newborns. In (b), the initial age distribution was 50 individuals in each age class. After some initial fluctuations, both populations settle into identical stable age distributions. On the logarithmic scale, the straight line for each age class indicates exponential increase.

number of individuals in each age class as a function of time. You can see that the graphs for the two populations initially appear quite different from one another as the relative numbers in the different age classes change in the early phases of population growth. In particular, you can see that the population with 200 newborns is dominated by this single age class, which passes as a cohort through the older age classes. However, after about 6 time steps, both populations have converged on the same age structure—they both have the same relative numbers in each age, with newborns being most common, and the oldest individuals being most rare. These relative proportions are maintained as the numbers in all ages increase exponentially.

These graphs illustrate an important property of age-structured populations. For most life tables, if a population is growing with constant birth and death rates, it will quickly converge on a **stable age distribution**, regardless of its initial age structure. In the stable age distribution, the *relative* numbers of individuals in each age class remain constant. Remember that the *absolute* numbers will increase exponentially, as evidenced by the linear population growth curves on the logarithmic scale of Figure 3.3. A special kind of stable age distribution is the **stationary age distribution**. In a stationary age distribution, $r = 0$, so both the relative and the absolute numbers in each age class remain constant.

What are the relative proportions in the different ages once the stable age distribution has been achieved? The proportion of the population represented by each age is just the number in that age divided by the total population size. This ratio is (Mertz 1970):

$$c(x) = \frac{e^{-rx}l(x)}{\sum\limits_{x=0}^{k} e^{-rx}l(x)} \qquad \text{Equation 3.12}$$

Once r has been calculated from the $l(x)$ and $b(x)$ schedules, Equation 3.12 can be used to determine the stable age distribution. The calculations are illustrated in Table 3.3. In a stable age distribution, newborns are the most common age, and the oldest age is least common. In most cases, the larger r is, the greater the proportion of the total population represented by newborns and young individuals. For the matrix algebra solution, the stable age distribution is the right-hand eigenvector of the Leslie matrix.

The Leslie matrix calculations of population growth can also be used as an independent check on the calculation of r. Table 3.4 illustrates some of the raw data of age structure and population size from Figure 3.3a. For any two consecutive time steps in the model, the ratio of the current population size to the previous population size is a measure of λ, the finite rate of increase. The final column of Table 3.4 gives the natural logarithm of this ratio, which is r.

Table 3.3 Calculation of stable age and reproductive value distributions.[a]

x	l(x)	b(x)	l(x)e⁻ʳˣ	c(x)	eʳˣ/l(x)	e⁻ʳʸl(y)b(y)	• e⁻ʳʸl(y)b(y)	v(x)
			Stable age distribution		**Reproductive value distribution**			
0	1.0	0	1.000	0.684	1.000	0.000	1.000	1.000
1	0.8	2	0.368	0.252	2.716	0.736	1.000	0.717
2	0.4	3	0.085	0.058	11.802	0.254	0.264	0.118
3	0.1	1	0.010	0.007	102.574	0.010	0.010	0.000
			• = 1.463					

Column headers: x, $l(x)$, $b(x)$, $l(x)e^{-rx}$, $c(x)$, $e^{rx}/l(x)$, $e^{-ry}l(y)b(y)$, $\bullet\ e^{-ry}l(y)b(y)$, $v(x)$. Stable age distribution spans $l(x)e^{-rx}$ and $c(x)$. Reproductive value distribution spans $e^{rx}/l(x)$ through $v(x)$.

[a] These calculations use $r = 0.776$, from the solution to the Euler equation in Table 3.1.

By 6 or 7 time steps in the model, the stable age distribution has been achieved, and the estimate of r is 0.776, which matches the calculation from the Euler equation in Table 3.1.

Table 3.4 Estimating r from the Leslie matrix calculations.[a]

Time step (t)	$n_1(t)$	$n_2(t)$	$n_3(t)$	$n_4(t)$	$n_{total}(t)$	$\lambda = \dfrac{n_{total}(t)}{n_{total}(t-1)}$	$r = ln(\lambda)$
0	200	0	0	0	200		
1	320	160	0	0	480	2.4	0.875
2	752	256	80	0	1088	2.267	0.818
.						.	.
.						.	.
.						.	.
6	16,549	6091	1402	161	24,203	2.173	0.776
7	35,965	13,239	3045	351	52,600	2.173	0.776
8	78,165	28,772	6620	761	114,318	2.173	0.776
.						.	.
.						.	.
.						.	.

[a] The data are from different time steps in Figure 3.3a. Fractions for the age-class values have been rounded to the nearest whole number.

Model Assumptions

In spite of the lengthy calculations, the model presented here shares the basic assumptions of the simple exponential growth model we derived in Chapter 1. In other words, we assume a closed population, no genetic structure, and no time lags. In the simple exponential model, we assumed that b and d were constant—they did not vary with time or with population density. In the age-structured model, we assume that the $l(x)$ and the $b(x)$ schedules are constant. As before, if each age class has a constant birth and death rate no matter how large the population, resources must be unlimited.

Incidentally, if we use the value of r from the Euler equation to forecast population growth, we must further assume that the population has achieved a stable age distribution. One final point is that we have described the $l(x)$ schedule from a cohort analysis, in which the fate of a cohort is followed through time. This **horizontal, or cohort life table** is the simplest method of obtaining the $l(x)$ schedule, but it assumes that death rates are constant during the time the cohort is followed. A more reliable method is to measure short-term death rates directly for each age class. Finally, it is possible to take a cross-section of the population at one time and estimate death rates from the relative sizes of consecutive age classes. This **vertical, or static life table** is much less reliable and assumes the population has reached a stationary age distribution. However, birth and death rates can be very difficult to measure in the field, and we often have to rely on a number of methods to piece together the data needed for a life-table analysis.

Model Variations

DERIVATION OF THE EULER EQUATION

The Euler equation forms the basis for age-structured demography, so it is important to understand how this equation is derived. The key to the Euler equation is recognizing the relationship between the number of births now and the number of births at some point in the past (Roughgarden 1979). The number of births in the population now, $B(t)$, is simply the sum of the number of births from parents of all different ages:

$$B(t) = \sum_{x=0}^{k} (\text{births from parents of age } x) \qquad \text{Expression 3.9}$$

If we allow the age intervals to become infinitely small, we can express this as an integral equation:

$$B(t) = \int_0^k (\text{births from parents of age } x) dx \qquad \text{Expression 3.10}$$

The number of births from parents of age x is the product of the number of individuals born at time $t - x$, their offspring production [$b(x)$], and their probability of surviving to age x [$l(x)$]:

$$B(t) = \int_0^k B(t-x)l(x)b(x)dx \qquad \text{Expression 3.11}$$

Remember that the number of births comes from a population that is increasing exponentially. Using C as an arbitrary starting population size, we have:

$$B(t) = Ce^{rt} \qquad \text{Expression 3.12}$$

Substituting this back into Expression 3.12 yields:

$$Ce^{rt} = \int_0^k Ce^{r(t-x)}l(x)b(x)dx \qquad \text{Expression 3.13}$$

Finally, if we divide both sides of Expression 3.14 by Ce^{rt}, we have the Euler equation:

$$1 = \int_0^k e^{-rx}l(x)b(x)dx \qquad \text{Equation 3.13}$$

As we noted earlier, the equivalent equation in discrete time is:

$$1 = \sum_{x=0}^k e^{-rx}l(x)b(x) \qquad \text{Equation 3.14}$$

REPRODUCTIVE VALUE

Using the Euler equation, we can calculate another useful statistic from the life table—the **reproductive value** of each age (Fisher 1930). The reproductive value is the relative number of offspring that remain to be born to individuals of a given age. You might think that a newborn individual would have the highest reproductive value because it has not yet produced any offspring. However, its reproductive value is discounted by the fact that it might not achieve its maximum potential lifespan and produce all of its potential offspring. Let $v(x)$ equal the reproductive value for an individual of age x. We can define reproductive value as the following ratio in a stable age distribution (Wilson and Bossert 1971):

$$v(x) = \frac{\text{number of offspring produced by individuals of age } x \text{ or older}}{\text{number of individuals of age } x} \qquad \text{Expression 3.14}$$

We can use the Euler equation to quantify the terms in the numerator and the denominator. For the numerator, we add the terms in the Euler equation from the current age forward:

$$\text{Offspring production} = \int_x^k e^{-ry}l(y)b(y)dy \quad \text{Expression 3.15}$$

For the denominator, the number of individuals in age x is the number born at time x in the past, multiplied by the probability of surviving to age x. Thus:

$$\text{Number in age } x = e^{-rx}\,l(x) \quad\quad \text{Expression 3.16}$$

Substituting Expressions 3.15 and 3.16 into 3.14 gives:

$$v(x) = \frac{\displaystyle\int_x^k e^{-ry}l(y)b(y)dy}{e^{-rx}l(x)} \quad\quad \text{Expression 3.17}$$

Rearranging the right-hand side yields a formula for reproductive value:

$$v(x) = \frac{e^{rx}}{l(x)}\int_x^k e^{-ry}l(y)b(y)dy \quad\quad \text{Equation 3.15}$$

The discrete-time version of Equation 3.15 allows us to use the $l(x)$ and $b(x)$ schedules to calculate the reproductive value for individuals of age x:

$$v(x) = \frac{e^{rx}}{l(x)}\sum_{y=x+1}^k e^{-ry}l(y)b(y) \quad\quad \text{Equation 3.16*}$$

For the matrix algebra solution, the left-hand eigenvector of the Leslie matrix is the vector of reproductive values. From Equation 3.15, the reproductive value of newborns always equals 1.0 ($v(0) = 1.0$). Thus, reproductive value is measured relative to that of the first age. For example, if $v(3) = 2.0$, an individual of age 3 will produce roughly twice as many offspring during the remainder of its lifetime as will a newborn. Reproductive value reflects the survivorship of an individual to its current age, its survivorship and reproduction in future ages, and the magnitude of r. Reproductive value usually peaks at or near the age of first reproduction, then drops off rapidly with later ages. For the data in Table 3.1, reproductive value is maximal for individuals of age 0 (Table 3.3).

---

*Be careful with the notation in this formula. In particular, notice that the summation subscript ($y = x + 1$) is increased by one. Thus, using the data from the sixth and eighth columns of Table 3.3, $v(1) = (2.716)(0.264) = 0.717$. Equation 3.16 generates reproductive values that are consistent with the matrix algebra solutions, but the formula is restricted to birth-pulse populations with a post-breeding census. See Goodman (1982) and Caswell (2001) for more details.

Reproductive value tells us which ages in the population are most "valuable" for future population growth. In Chapter 2, we noted that maximum yield for a harvested population occurred when the population was harvested to maximize population growth rate. For the simple logistic model, the best strategy turned out to be maintaining the population at $K/2$. For an age-structured population, maximizing population growth rate would mean harvesting individuals with relatively low reproductive value—usually newborns and very old individuals, depending on the age structure of the population.

Reproductive value is also relevant to problems of population management and conservation biology. If we are going to transplant captive-bred individuals to a new population in order to increase the population growth rate, we should wait until those individuals reach the age with the highest reproductive value. Finally, natural selection will operate most heavily on ages with high reproductive value. For example, an allele that expresses deleterious effects in reproductive age classes will be eliminated by selection much more quickly than an allele that expresses the effects in older age classes, with lower reproductive value. **Senescence** may represent the accumulation of deleterious effects in old individuals. Selection pressure is weaker on older individuals (Rose 1984), in part because of their lower reproductive value (Fisher 1930).

LIFE HISTORY STRATEGIES

Life-table data are essential for ecological predictions of population growth rates and age structure. From an evolutionary perspective, we can ask why we see certain life history patterns. In other words, why has natural selection favored certain $l(x)$ and $b(x)$ schedules? Selection will favor any life history schedule that maximizes an individual's contribution of offspring to the next generation. Thus, the "perfect" life history schedule would be one with maximum survivorship and maximum fertility in all age classes!

However, two forces prevent the evolution of this optimal life history. First, we expect a number of **tradeoffs** to occur among life history traits. Organisms that invest heavily in reproduction have less energy to devote towards growth, maintenance, and resource acquisition. This may lead to tradeoffs between reproduction and survivorship. An organism may produce many small offspring that survive poorly or a few large offspring that survive well. Hence, there may be tradeoffs between offspring number and offspring survivorship.

Life history strategies will also be shaped by **constraints**—physiological or evolutionary limitations that prevent the evolution of certain life history traits. For example, organisms with large body size must take longer to grow and reach maturity, so the age at first reproduction may be constrained by body size. If an organism bears live offspring, body size will also constrain the number of offspring produced. The life history traits of an organism may reflect a long evolutionary heritage, and may not represent the best "solution" to the problem of maximizing fitness in the organism's current environment.

One popular body of theory envisions that relative population density serves as an important selective force on life history traits (MacArthur and Wilson 1967; Pianka 1970). The theory of **r–K selection** takes its name from the two constants of the logistic growth equation. Imagine a population that is maintained at low population density, so that resources for growth are not limited. Under these circumstances, the best reproductive strategy is simply to maximize offspring production. So, the traits expected under r-selection are early, semelparous reproduction, large r, many offspring with poor survivorship, a Type III survivorship curve, and small adult body size.

By contrast, in K-selection, an organism is growing in an environment that is chronically crowded. An r-strategy will not work in this case because the offspring will face limited resources and be relatively poor competitors. Instead, the best strategy is one that leads to fewer, high-quality offspring that are superior competitors. With resource limitation, K-selection should favor late, iteroparous reproduction, small r, few offspring with good survivorship, a Type I survivorship curve, and large adult body size. Classic examples of species thought to have evolved under the different regimes include mosquitoes and weeds (r-selected), and humans and whales (K-selected).

In spite of its popularity in textbooks, the theory of r- and K-selection is beset by a number of problems. One fundamental problem is that the "predictions" of r–K selection theory were never derived from a population model with age structure. Another difficulty is that population density is not the only force driving the evolution of life history traits. For example, the theory predicts that iteroparity evolves when organisms face resource competition and must devote more of their energy to growth and maintenance than to reproduction. But iteroparity could also evolve as a "bet-hedging" strategy if the survival of offspring is uncertain from one time period to the next (Murphy 1968). It may be advantageous to spread reproduction over many time periods if there is a risk of losing all your offspring if they are born at the wrong time.

Moreover, not all organisms have life history traits that neatly fit the predictions of the model. For example, many forest trees are long-lived and iteroparous (K-selection), but they have a Type III survivorship curve (r-selection). Finally, the r–K selection theory has not been confirmed experimentally. Laboratory populations of fruit flies (Taylor and Condra 1980) and protozoa (Luckinbill 1979) did not always evolve r-selected traits when they were maintained in uncrowded conditions or K-selected traits when they were maintained in crowded conditions. Although the original theory of r–K selection has been discarded, it is nevertheless true that the ecological conditions an organism experiences—including its population density—can be important forces of natural selection that shape life histories.

For example, mortality from predators can lead to major changes in birth and death schedules (Gadgil and Bossert 1970; Roff 1992). If predators specialize on adult size classes, natural selection will favor individuals that mature early and reproduce at small body sizes. These predictions have been confirmed for freshwater tropical guppies: life histories differ among populations of the same species, depending on whether or not predators are present (Reznick et al. 1996). Moreover, field studies demonstrate that life history traits can evolve very rapidly in response to the presence of predators (Reznick et al. 1997). Other studies have shown that body size—and hence some life history traits—can also evolve in response to the presence of competing species (Schluter 1994). In Chapters 5 and 6, we will develop ecological models for understanding the effects of predators and competitors on population dynamics. But it is important to emphasize that these interactions have consequences for the evolution of life histories as well.

STAGE- AND SIZE-STRUCTURED POPULATION GROWTH

An implicit assumption in our development of the life table model is that the age of an organism is the "correct" variable to use in defining the life history. But for many life histories, age is not the critical variable. For example, many insects pass through egg, larval, pupal, and adult stages. Survival may be influenced more by an insect's stage than its age. That is, survival of a beetle may not depend on whether the beetle is three or six months old, but on whether it is in the larval or adult stage. Of course, age and stage are not independent of one another, because an organism's life history stage will depend, in part, on how old it is. But the transitions between stages are often flexible and depend on biotic factors, such as food supply and population density, and abiotic factors, such as temperature and photoperiod.

Even for organisms that do not have distinct life history stages, survival and reproduction may depend more on the size or an organism than on its age. Many organisms have indeterminate growth—a small fish may be either a fast-growing juvenile or a stunted adult. If the risk of mortality is from predation by other fishes, only the individual's size, rather than its age, may be relevant. Finally, "modular" organisms such as plants and corals may be organized as colonies or semi-independent units (plant shoots) that are capable of reproduction. In these cases, the life history may be extremely complex, as coral colonies can fragment or fuse, and plants can reproduce through vegetative propagation. In all these examples, the age of the organism is less important than its size or stage in determining its survivorship and reproduction.

Fortunately, the Leslie matrix can be modified to account for these kinds of life histories (Lefkovitch 1965). The key change is that the entries in the population matrix no longer represent the age of an organism, but rather its stage (or size). We still incorporate a time step that represents the transition from one stage to the next. For example, here is a transition matrix for a simplified insect life cycle with three stages—egg, larva, and adult:

$$
\begin{array}{c}
\\
\text{egg} \\
\text{larva} \\
\text{adult}
\end{array}
\begin{array}{ccc}
\text{egg} & \text{larva} & \text{adult} \\
\left[\begin{array}{ccc}
0 & 0 & F_{ae} \\
P_{el} & P_{ll} & 0 \\
0 & P_{la} & P_{aa}
\end{array}\right]
\end{array}
\qquad \text{Expression 3.18}
$$

Remember that each column represents the stage at time t and each row represents the stage at time $t + 1$. The entries in the first row represent fertilities. The entries in the other rows represent transition probabilities between stages. In contrast to the Leslie matrix, we now have positive entries on the diagonal. This means that larvae and adults can stay in a particular stage at a given time, whereas eggs will either die or advance to the larval stage. Only the adult can reproduce, so there is a single fertility entry (F_{ae}) for this stage.

Here is a transition matrix for a long-lived forest tree that is classified into five size classes:

$$
\begin{array}{c}
\\
\text{size 1} \\
\text{size 2} \\
\text{size 3} \\
\text{size 4} \\
\text{size 5}
\end{array}
\begin{array}{ccccc}
\text{size 1} & \text{size 2} & \text{size 3} & \text{size 4} & \text{size 5} \\
\left[\begin{array}{ccccc}
P_{11} & F_{21} & F_{31} & F_{41} & F_{51} \\
P_{12} & P_{22} & 0 & 0 & 0 \\
0 & P_{23} & P_{33} & 0 & 0 \\
0 & 0 & P_{34} & P_{44} & 0 \\
0 & 0 & 0 & P_{45} & P_{55}
\end{array}\right]
\end{array}
\qquad \text{Expression 3.19}
$$

Again, there is the possibility that an individual will remain in the same size class (diagonal elements) or grow to the next consecutive size class (subdiagonal elements). All size classes except the first reproduce, giving positive fertility values in the first row of the matrix.

As a final, and more complex, example, consider a population of reef-building corals with three size classes (small, medium, and large):

$$
\begin{array}{c}
\\
\text{small} \\
\text{medium} \\
\text{large}
\end{array}
\begin{array}{ccc}
\text{small} & \text{medium} & \text{large} \\
\left[\begin{array}{ccc}
P_{ss} + P_{ss} & P_{ms} + F_{ms} & P_{ls} + F_{ls} \\
P_{sm} & P_{mm} & P_{lm} \\
P_{sl} & P_{ml} & P_{ll}
\end{array}\right]
\end{array}
\qquad \text{Expression 3.20}
$$

As before, the diagonal elements represent the probability that a colony remains in the same size class, and the subdiagonal elements represent the probability that a colony grows to the next size class. However, there is now the possibility that large colonies can fragment into medium (P_{lm}) or small (P_{ls}) colonies, and that medium colonies can fragment into small colonies (P_{ms}). Small colonies can also fuse with one another, thus "skipping" a stage and going directly from small to large (P_{sl}). Finally, look at the first row of the matrix and notice that the entries are sums of fecundities and stage transitions. This relationship occurs because the production of small colonies has components of sexual reproduction (F) and asexual fragmentation and persistence (P).

As illustrated in Figure 3.4, these complex life cycles can also be represented in loop diagrams. Each circle in the loop represents a different life his-

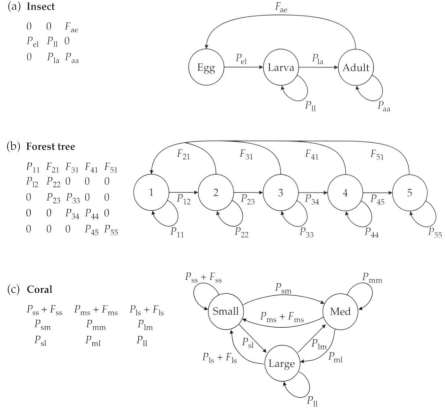

Figure 3.4 Stage-transition matrices and loop diagrams for different life histories. (a) Simplified insect life history. (b) Long-lived forest tree life history. (c) Coral life history, with sexual and asexual reproduction.

tory stage, and each arrow represents a transition from one stage to the next. Stages not connected by arrows have a zero for the corresponding entry in the transition matrix.

In spite of the complexities of these life cycles, the mechanics of the matrix multiplication are exactly the same as for the simple Leslie matrix. As long as the transition elements are constant, the population will eventually exhibit exponential growth and a stable stage distribution. However, we can no longer use the Euler equation for these life histories, and must obtain the matrix solutions for r and the stable stage distribution. For any transition matrix, λ is the dominant eigenvalue. The right-hand eigenvector is the stable stage distribution, and the left-hand eigenvector is the reproductive value distribution (Caswell 1989). The matrix approach allows us to use the same analytical framework to study complex life histories that do not fit a simple age classification.

Empirical Examples

LIFE TABLES FOR GROUND SQUIRRELS

A long-term demographic study of the Uinta ground squirrel (*Spermophilus armatus*) demonstrates the importance of life-table analysis in understanding population growth (Slade and Balph 1974). At a field station in northern Utah, squirrels emerged from hibernation each year between late March and mid-April, depending on the weather. Females bred shortly after they emerged and established territories. The first young were born in early May, and juveniles left their natal burrows about three weeks later. During June and July, all age classes and sexes in the population were active. Adults began hibernating in July, and by September all squirrels had disappeared underground.

Researchers trapped and tagged all individuals in the 8.9-hectare study area and monitored their activity from observation towers. The research was conducted over a seven-year period and divided into two phases. During the first phase (1964–1968), the population was left undisturbed, except for the monitoring. Population size fluctuated from 178 to 255, with a mean of 205. During the second phase (1968–1971), researchers reduced the squirrel population to about 100 individuals. Life-table analysis (Table 3.5) revealed the dramatic effects of density reduction on growth rate and age structure.

Before the population was reduced, age-specific birth and death rates were approximately balanced, generating a slightly negative growth rate [r = −0.046 individuals/(individual • year)]. The maximum life span was approximately five years, although this varied somewhat between different habitats. In the stable age distribution, 37% of the population was juveniles (Figure 3.5a), and reproductive value peaked for individuals during their second year (Figure 3.5b).

Table 3.5 Life tables for Uinta ground squirrels (*Spermophilus armatus*) before and after density reduction.

	Pre-reduction life table		Post-reduction life table	
x (years)	l(x)	b(x)	l(x)	b(x)
0.00	1.000	0.00	1.000	0.00
0.25	0.662	0.00	0.783	0.00
0.75	0.332	1.29	0.398	1.71
1.25	0.251	0.00	0.288	0.00
1.75	0.142	2.08	0.211	2.24
2.25	0.104	0.00	0.167	0.00
2.75	0.061	2.08	0.115	2.24
3.75	0.026	2.08	0.060	2.24
4.75	0.011	2.08	0.034	2.24
5.75	0.000	0.00	0.019	2.24
6.75	—	—	0.010	2.24
7.75	—	—	0.000	0.00

Data from Slade and Balph (1974).

After density was reduced, reproduction exceeded mortality, and there was a substantial rate of population increase [$r = 0.306$ individuals/(individual • year)]. The maximum life span increased to seven years, and the stable age distribution shifted slightly toward older ages (Figure 3.5a). The reproductive value showed a broader peak for three- and four-year olds (Figure 3.5b), reflecting the increased reproduction and survival of older ages.

The density reductions revealed that crowding had many effects beyond a slowing of population growth rate. Survivorship, reproduction, life span, and age structure were all sensitive to population density. The manipulations also point to a key weakness of our exponential growth model: age-specific birth and death rates change with population size!

Density dependence can be incorporated into either the mortality or the fecundity schedules for one or more age classes. Even if it limits the increase of only a single age class, density dependence can be an effective brake on total population growth, and can lead to complex population dynamics. In the remainder of this primer, we will return to simple models of populations that do not incorporate age structure. However, the biological details of migration (Chapter 4), competition (Chapter 5), predation (Chapter 6), and colonization (Chapter 7) almost certainly reflect the age and size structure within a population.

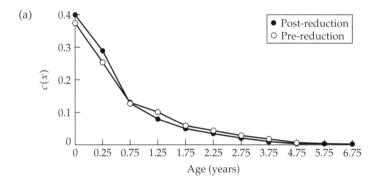

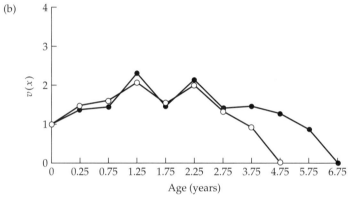

Figure 3.5 (a) Stable age distribution and (b) reproductive value distribution for Uinta ground squirrels (*Spermophilus armatus*) before and after density reduction. Data from Table 3.5.

STAGE PROJECTION MATRICES FOR TEASEL

Teasel (*Dipsacus sylvestris*) is a European perennial "weed" that is common in abandoned fields and meadows of the eastern United States. The plant has a complex life cycle that can be described with a stage-based matrix model. Most seeds fall within two meters of the adult plant, and the seeds may lie dormant for one or two years. Seeds that successfully germinate form a large-leafed rosette. The rosette phase is variable and may last for more than five years. The rosette requires cold-hardening (vernalization) before it will form a flowering stalk the following summer. Teasel flowers and sets seed only once, and then the plant dies.

Teasel was studied in eight abandoned fields in Michigan, which were sown with teasel seed at the start of the study (Werner 1977; Werner and Caswell 1977). To construct the stage-based transition matrix, individual plants were monitored in marked plots for several consecutive years. The life cycle of teasel can be divided into six stages (Caswell 1989):

1. Dormant first-year seeds
2. Dormant second-year seeds
3. Small rosettes (<2.5 cm diameter)
4. Medium rosettes (2.5–18.9 cm diameter)
5. Large rosettes (≥19.0 cm diameter)
6. Flowering plants

Figure 3.6 gives the loop diagram and corresponding stage matrix for this life cycle as measured on one of the eight experimental plots. From the positive entries on the diagonals and subdiagonals, the rosettes can remain in their own size class, grow to a larger rosette size, or flower. The single entry in the first

Seed (1)	Seed (2)	Ros (s)	Ros (m)	Ros (l)	Flowering plant
0	0	0	0	0	322.380
0.966	0	0	0	0	0
0.013	0.010	0.125	0	0	3.448
0.007	0	0.125	0.238	0	30.170
0.008	0	0	0.245	0.167	0.862
0	0	0	0.023	0.750	0

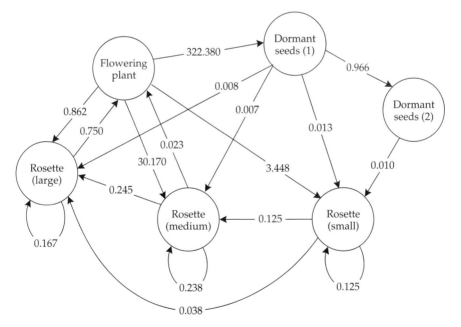

Figure 3.6 Transition matrix and loop diagram for teasel (*Dipsacus sylvestris*). Transitions are shown for dormant first-year and second-year seeds [seed (1) and seed (2)], small, medium, and large rosettes [ros (s), ros (m), ros (l)], and flowering plants. (Data from Caswell 1989.)

row of the matrix reflects the fact that only the flowering plants can produce seed. Also, notice that the diagonal element is zero for flowering plants (P_{66}), indicating that they do not survive after they flower. The population growth rate for this matrix is $\lambda = 2.3242$. This corresponds to an r of 0.8434 individuals / (individual • year), with a projected doubling time of less than 10 months.

In contrast to a simple age-classified model, relative frequencies in the stable stage distribution do not always decrease with later stages. In the stable stage distribution for teasel, there were more medium than small rosettes (Figure 3.7a). Reproductive values varied over six orders of magnitude, from a minimum for second-year dormant seeds to a maximum for flowering plants (Figure 3.7b).

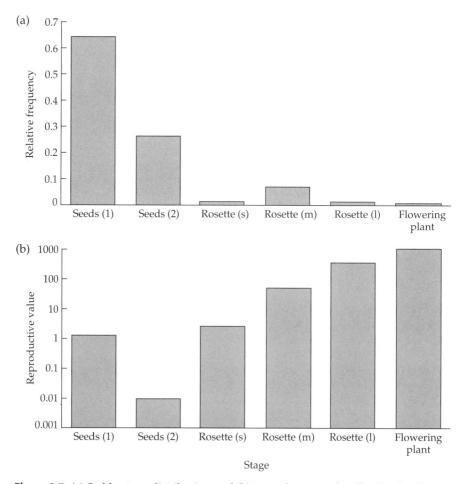

Figure 3.7 (a) Stable-stage distribution and (b) reproductive-value distribution for teasel (*Dipsacus sylvestris*). Note logarithmic scale. Derived from data in Figure 3.6.

These same data were also analyzed as an age-classified model, treating rosettes of 1–4 years as separate age classes (Werner and Caswell 1977). However, the stage-based model predicted the year of first flowering more accurately than did the age-based model. The results suggested that the size of a rosette, rather than its age, is the more important determinant of growth and survivorship for teasel.

The model results for teasel varied greatly between different fields, and population growth rates (r) ranged from −0.46 to 0.96 individuals / (individual • year). Fields with the lowest r had high levels of grass litter, which suppressed teasel seed germination. Population growth rate was also low in fields with high densities of herbaceous plants, which reduced survivorship of teasel rosettes through competition and shading. Finally, r was correlated with annual primary productivity of a field. Population growth rates were highest in the least productive fields, perhaps reflecting competition with other plants. The very high rates of increase measured for some teasel populations are unlikely to be sustained in the long run. As in the ground squirrel example, a density-dependent model may be more appropriate for forecasting population size.

Problems

3.1. Plot the logarithm (base 10) of squirrel survivorship for the pre- and post-reduction populations (Table 3.5). What is the general shape of these curves (Type I, II, or III), and how does density reduction affect survivorship?

3.2. Here is a set of hypothetical life-table data for a population of snails:

Age in years (x)	S(x)	b(x)
0	500	0.0
1	400	2.5
2	40	3.0
3	0	0.0

a. Complete the life-table analysis by calculating $l(x)$, $g(x)$, R_0, G, and the estimate of r. Calculate the exact value of r with the Euler equation.

b. Determine the stable age and reproductive value distributions for this life table.

*3.3. Suppose the snail population in Problem 3.2 consisted of 50 newborns, 100 one-year-olds, and 20 two-year-olds. Construct the Leslie matrix for this life table, and project population growth for the next two consecutive years.

* Advanced problem

CHAPTER 4

Metapopulations

Model Presentation and Predictions

In Chapters 1–3, we explored several models of population growth. These models differed in their major assumptions: unlimited resources (Chapter 1) versus a finite carrying capacity (Chapter 2), and a homogenous population (Chapter 1) versus an age-structured one (Chapter 3). All of these models described a **closed population**. In other words, the population changed size because of births and deaths that occurred locally. We explicitly assumed that individuals did not move between populations.

This assumption of a closed population was mathematically convenient, but not biologically realistic. For migratory animals, such as North American songbirds that overwinter in the tropics, or oceanic salmon that spawn in freshwater streams, the seasonal movement of individuals is the dominant cause of population change. Many nonmigratory species also move between populations. In particular, organisms with complex life histories often have seeds or larvae that are adapted for movement to new populations. The ascidians described in Chapter 2 are a good example. The adults are filter-feeding invertebrates that attach permanently to rock walls, but the "tadpole" larvae are free-swimming and drift in the current for several days before settlement and metamorphosis. Consequently, the "births" in a local ascidian population consist of juveniles that originated from many different sites.

The movement of individuals between populations may be density-dependent. In territorial species, such as black-throated blue warblers (*Dendroica caerulescens*), not all individuals are able to establish territories, and those that do not may migrate in search of less crowded populations. Mathematical models that ignore the biology of animal and plant movement may not give an accurate description of population dynamics.

In this chapter, we will develop a class of simple models that takes into account the fact that individuals do move among sites and that such movement is potentially important to the persistence and survival of populations. This chapter explores the concept of a **metapopulation**. The metapopulation can be thought of as a "population of populations" (Levins 1970)—a group of several local populations that are linked by immigration and emigration.

In order to study metapopulations, we need to make two important shifts in our frame of reference. The first shift concerns how we measure populations. In Chapters 1–3, our models predicted the *size* of a population—the number of individuals at equilibrium. Our metapopulation models will not predict the size of a population, but only its *persistence*. Thus, the range of numbers representing population size will be collapsed to only two possible values: 0 (local extinction) or 1 (local persistence). We will no longer distinguish between large and small populations, or between populations that cycle, fluctuate, or remain constant. Instead, the only distinction is between populations that persist and those that go extinct.

The second shift concerns the spatial scale at which we study populations. In Chapters 1–3 we emphasized equilibrium solutions for population size, implicitly focusing on local populations that persisted through time. In contrast, the metapopulation perspective is that local populations frequently go extinct, and that the appropriate spatial scale for recognizing an equilibrium is the regional or landscape level, which encompasses many connected sites. At this scale, we will no longer focus on the persistence of any particular population. Instead, we will build a model that describes the fraction of all population sites that are occupied. Thus, we will ignore the fate of individual populations and model the extent to which populations fill the landscape. This large-scale view will allow us to use simple mathematics and avoid the complexities of trying to explicitly model local population size and individual migration. As an analogy, if we were modeling the dynamics of a busy parking lot, we would try to predict how many parking spaces were filled, not which particular spaces were occupied.

METAPOPULATIONS AND EXTINCTION RISK

The metapopulation perspective allows us to make a distinction between **local extinction**, in which a single population disappears, and **regional extinction**, in which all populations in the system die out. Even if populations are not connected by migration, the risk of regional extinction is usually much lower than the risk of local extinction.

To explore this concept quantitatively, we can define p_e as the **probability of local extinction**—that is, the probability that the population in an occupied patch goes extinct. This probability is a number that ranges between 0 and 1. If $p_e = 0$, persistence is certain, whereas if $p_e = 1$, extinction is certain. All populations go extinct in the long run, so probabilities of extinction must be measured relative to a particular time scale. For metapopulation dynamics, the appropriate time scale is often years or decades.

Suppose that $p_e = 0.7$, for probabilities measured on a yearly time scale. This means there is a 70% chance (100×0.7) that a population will go extinct during a single year, and a 30% chance that it will persist ($1 - p_e = 0.3$). What are the chances that the population will persist for two years? The **probability of persistence** for two years is the probability of no extinction in the first year $(1 - p_e)$ multiplied by the probability of no extinction in the second year $(1 - p_e)$. Thus:

$$P_2 = (1 - p_e)(1 - p_e) = (1 - p_e)^2 \qquad \text{Expression 4.1}$$

The probability that a population will persist for n years (P_n) is the probability of no extinction for n years in a row:

$$P_n = (1 - p_e)^n \qquad \text{Equation 4.1}$$

For example, if $p_e = 0.7$, and $n = 5$, $P_n = (1 - 0.7)^5 = 0.00243$. So, if there is a 70% chance that a population goes extinct in one year, the chance of persistence for five years in a row is only 0.2%.

Now suppose that instead of a single population, we have two identical populations, each with a p_e of 0.7. For now, we assume that these populations are independent of one another—the chance of extinction in one patch is not affected by the presence or absence of populations in other patches. For this pair of populations, what is the probability of regional persistence, that is, what are the chances that *at least* one population persists for one year? The probability of regional persistence for one year (P_x) is 1 minus the probability that both patches go extinct during the year:

$$P_2 = 1 - (p_e)(p_e) = 1 - (p_e)^2 \qquad \text{Expression 4.2}$$

The **probability of regional persistence** in a set of x patches is the probability that all x patches do not go simultaneously extinct:

$$P_x = 1 - (p_e)^x \qquad \text{Equation 4.2}$$

Thus, if we had 10 patches, each with $p_e = 0.7$, the probability of regional persistence is $P_{10} = 1 - (0.7)^{10} = 0.97$. In other words, with 10 patches, there is a 97% chance that at least one population will persist, even though it is likely that any particular population will go extinct ($p_e = 0.7$)! Figure 4.1 shows that P_x increases rapidly as more patches are added, although there is an overall decrease as p_e is increased.

Equation 4.2 illustrates an important principle: multiple patches "spread the risk" of extinction. Even if individual populations are doomed to extinction, a set of populations can persist for a surprisingly long time. In the next section, we will build metapopulation models in which these local populations are linked to one another, so that probabilities of local extinction and local colonization depend on patch occupancy.

A MODEL OF METAPOPULATION DYNAMICS

Imagine a set of homogenous patches, each of which can be occupied by a single population. Let f equal the **fraction of sites occupied**, that is, the proportion of patches that contain populations. Thus, f is a number constrained between 0 and 1. If $f = 1$, all sites are occupied by populations, and the landscape is saturated. If $f = 0$, all sites are unoccupied, and the metapopulation is regionally extinct.

How does f change with time? f can increase if empty sites are successfully colonized. Let I = the **immigration rate**: the proportion of sites successfully

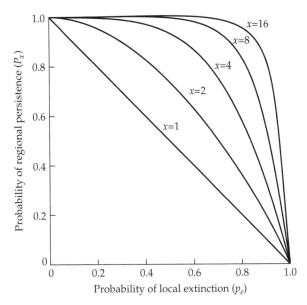

Figure 4.1 The relationship between the probability of regional persistence (P_x), the probability of local extinction (p_e), and the number of populations (x). Note that as the number of populations is increased, the probability of regional persistence is substantially higher, for a given probability of local extinction.

colonized per unit time.[*] f can also decrease if occupied sites undergo extinction. Let E = the **extinction rate**, that is, the proportion of sites that go extinct per unit time. The change in f is determined by the balance of gains from colonization and losses from extinction:

$$\frac{df}{dt} = I - E \qquad \text{Expression 4.3}$$

There is a close analogy between Expression 4.3 and our initial derivation of the exponential growth model in Chapter 1 (Expression 1.5). In the exponential growth model, there was continuous turnover of individuals from births and deaths. Population size (N) reached an equilibrium only if the birth rate precisely equaled the death rate. Similarly, at the metapopulation level, there is continuous turnover of individual *populations* through colonization and extinction. The fraction of population sites (f) reaches an equilibrium when the immigration rate precisely equals the extinction rate. We will see this same derivation once more in Chapter 7, when we model the number of species in a community.

---

[*]Technically, we should refer to this as the colonization, not the immigration, rate, but we use this terminology for consistency with the MacArthur–Wilson model, which is developed in Chapter 7.

We now wish to describe the immigration and extinction functions in more detail. The immigration rate depends on two factors. First is the **probability of local colonization**, p_i. If each site is colonized independently, this probability will depend only on the physical and biological conditions within a patch. Many factors can affect p_i, including patch area, the availability of critical habitats and food resources, and the absence or scarcity of predators, pathogens, and competitors.

The probability of local colonization can also be affected by factors that are external to the site. Specifically, if the sites are linked by migration, the probability of colonization may depend on the presence of populations in other sites. In other words, when many sites are occupied (large f), there are many individuals migrating, so the probability of colonization is higher than when few sites are occupied (small f). Therefore, p_i will depend on f. In the following sections, we will develop models in which p_i is either dependent or independent of f (Gotelli 1991).

The immigration rate depends not only on p_i, but also on the availability of unoccupied sites, which is measured by $(1 - f)$. The more sites available for colonization, the faster the overall immigration rate. Thus, the immigration rate is the product of the probability of local colonization (p_i) and the fraction of unoccupied sites $(1 - f)$:

$$I = p_i(1 - f) \qquad \text{Expression 4.4}$$

The immigration rate will equal zero in two cases: first, if the probability of local colonization is zero $(p_i = 0)$; and second, if all the sites in the metapopulation are occupied $(f = 1)$.

If we follow a similar line of reasoning, the extinction rate, E, is the product of the probability of local extinction (p_e) and the fraction of sites occupied (f):

$$E = p_e f \qquad \text{Expression 4.5}$$

The extinction rate equals zero if the probability of extinction is zero $(p_e = 0)$, or if none of the sites in the metapopulation is occupied $(f = 0)$. Substituting Expressions 4.4 and 4.5 back into 4.3 gives us a general metapopulation model:

$$\frac{df}{dt} = p_i(1 - f) - p_e f \qquad \text{Equation 4.3*}$$

---

*Because this is a continuous differential equation, p_i and p_e are technically not probabilities, but fractional rates. However, p_i and p_e behave as probabilities when they are multiplied by a finite time interval. Over such a time interval, we would need to add a correction term to Equation 4.3 to account for the chance that an occupied patch could undergo an extinction and a recolonization (Ray et al. 1991). However, the correction term is small, and it is simpler to use the continuous differential equation and to interpret p_i and p_e as immigration and extinction probabilities.

Equation 4.3 is a simple model of metapopulation dynamics that will serve as a template for developing alternative models. By changing some of our assumptions about colonization and extinction processes, we can generate new metapopulation models that make different predictions about the fraction of sites occupied at equilibrium ($\hat{f}$). Before exploring these variations, we will first examine the general assumptions of this model.

Model Assumptions

Equation 4.3 makes the following assumptions:

✔ *Homogenous patches.* The population sites must not differ in their size, isolation, habitat quality, resource levels, or other factors that would affect the probability of local colonization and local extinction.

✔ *No spatial structure.* The model assumes that probabilities of colonization and extinction may be affected by the overall fraction of occupied sites (f), but not their spatial arrangement. In a more realistic metapopulation model, the probability of colonization for a particular site would depend on the occupancy of close neighboring patches, rather than on the overall f. This sort of "neighborhood" model can be studied by computer simulation or by using equations of diffusion, in which the spread of populations through empty sites is analogous to the dispersion of an ink droplet through a beaker of water.

✔ *No time lags.* Because we are describing metapopulation dynamics with a continuous differential equation, we assume that the metapopulation "growth rate" (df/dt) responds instantly to changes in f, p_i, or p_e.

✔ *Constant p_e and p_i.* The probabilities p_e and p_i do not change from one time period to the next. Although we cannot say precisely which populations will go extinct and which will be colonized, the probabilities of these events do not change.

✔ *Regional occurrence (f) affects local colonization (p_i) and extinction (p_e).* Except for the basic island–mainland model (see below), metapopulation models assume that migration is substantial enough to affect local population dynamics and influence probabilities of colonization and/or extinction. Consequently, p_i and/or p_e are functions of f.

✔ *Large number of patches.* The fraction of occupied sites in our model can become infinitely small, and the metapopulation will still persist. Thus, we are not assuming any demographic stochasticity (see Chapter 1) of the metapopulation due to small patch numbers.

Model Variations

THE ISLAND–MAINLAND MODEL

The simplest model for our metapopulation is that both the p_i and p_e are constants. If p_e is a constant, the probability of extinction is the same for each population and does not depend on the fraction of patches occupied. This assumption is analogous to a density-independent death rate in a population growth model, because the death rate does not depend on population size (see Chapter 2). Similarly, the probability of colonization may be fixed. Constant p_i implies a **propagule rain**—a continuous source of migrants that could potentially colonize an empty site (Figure 4.2a). If there is a large, stable "mainland" population, it may generate a propagule rain for a set of "islands" in the metapopulation. A propagule rain may also characterize some plant populations that may be colonized by a seed bank of long-lived buried seeds. The equilibrium value of f for this **island–mainland model** can be found by setting Equation 4.3 equal to zero and solving for f:

$$0 = p_i - p_i f - p_e f \qquad \text{Expression 4.6}$$

$$p_i f + p_e f = p_i \qquad \text{Expression 4.7}$$

Dividing both sides of Expression 4.7 by $(p_i + p_e)$ gives $\hat{f}$, the equilibrium for f:

$$\hat{f} = \frac{p_i}{p_i + p_e} \qquad \text{Equation 4.4}$$

In the island–mainland model, the fraction of sites occupied at equilibrium is a balance between extinction and immigration probabilities. Notice that even if the probability of extinction (p_e) is very large and the probability of colonization (p_i) is very small, at least some of the sites in the metapopulation will be occupied ($\hat{f} > 0$), because the metapopulation is continually replenished by the external propagule rain.

INTERNAL COLONIZATION

Now we will relax the assumption of the propagule rain and instead imagine that the only source of propagules for the metapopulation is the set of occupied population sites (Figure 4.2b). In other words, there is **internal colonization** such that:

$$p_i = if \qquad \text{Expression 4.8}$$

The constant i is a measure of how much the probability of colonization of empty sites increases with each additional patch that is occupied. In this

(a)

(b)

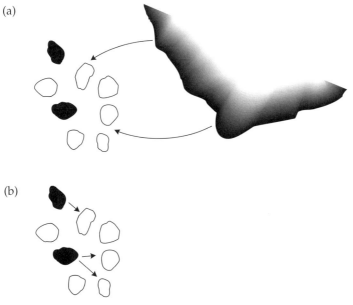

Figure 4.2 (a) Colonization in the island–mainland model. Colonists for a set of islands always come from a large mainland area. Open islands represent empty sites and filled islands represent sites that contain populations. (b) Colonization in the internal model. Colonists do not originate from a permanent external source, but instead originate only from currently occupied islands.

model, each population contributes individuals to a pool of propagules, which then have the potential to colonize unoccupied sites. Note that if all of the populations go extinct ($f = 0$), the probability of colonization goes to zero because there is no other source of colonists. This condition is in contrast to the island–mainland model, in which colonists were always present because of the external mainland population.

Assuming that the extinctions are still independent and substituting Expression 4.8 back into the general model (Equation 4.3) gives (Levins 1970):

$$\frac{df}{dt} = if(1-f) - p_e f \qquad \text{Equation 4.5}$$

Again, we set this equation equal to zero and solve for the equilibrium f:

$$p_e f = if(1-f) \qquad \text{Expression 4.9}$$

$$p_e = i - if \qquad \text{Expression 4.10}$$

$$if = i - p_e \qquad \text{Expression 4.11}$$

Dividing both sides by i yields:

$$\hat{f} = 1 - \frac{p_e}{i}$$

Equation 4.6

In contrast to the predictions of the island–mainland model, persistence of the metapopulation ($\hat{f} > 0$) is no longer guaranteed. Instead, the metapopulation will persist only if the strength of the internal colonization effect (i) is greater than the probability of local extinction (p_e). If this condition is not met, the metapopulation will go extinct ($\hat{f} \le 0$). Extinction can happen because the metapopulation is no longer receiving the benefit of external colonization.

THE RESCUE EFFECT

Our first two metapopulation models (island–mainland and internal colonization) both assumed that the probability of extinction was independent of the fraction of sites occupied. Now we should consider the possibility that extinction might be affected by f. How might this happen? As before, we assume that each occupied site produces an excess number of propagules that leave the site and arrive at other populations. If the propagules arrive at an empty site, they represent potential colonists. If conditions are good, these propagules may be able to establish a breeding population in the site. But migrants may also arrive at occupied sites and increase the size of established populations. This increase in N is a **rescue effect** that may prevent the local population from going extinct due to demographic and environmental stochasticity (see Chapter 1). The rescue effect is defined as the reduction in the probability of extinction that occurs when more population sites are occupied, and hence more individuals are available to boost local population sizes.

The tradeoff of propagules leaving a site and those entering from other sites cannot be strictly linear. Otherwise, it would not be possible to achieve a rescue effect—the reduction in p_e due to immigration would be canceled by the increase in p_e due to emigration. However, the loss of some individuals as migrants may have a negligible effect on p_e. In fact, if migration is density-dependent, individuals that do not migrate to other population sites might reproduce or survive poorly in their sites of origin. An explicit model of the rescue effect would need to include parameters for N, p_e, and migration. But we can capture the essence of the rescue effect in our simple metapopulation model by assuming that:

$$p_e = e(1 - f)$$

Expression 4.12

Expression 4.12 says that the probability of local extinction decreases as more population sites are occupied. e is a measure of the strength of the rescue effect, because it controls how much p_e decreases with the addition of another occupied site. Notice that if all population sites are occupied ($f = 1$), the probability of local extinction is zero. This is unrealistic, because even in a saturated landscape there should be some intrinsic background extinction risk. But we would have to introduce another parameter into the model to account for background extinction, so instead we will use Expression 4.12 to keep things simple. Assuming an external propagule rain and a rescue effect, we substitute Expression 4.12 into the general model (Equation 4.3):

$$\frac{df}{dt} = p_i(1-f) - ef(1-f) \qquad \text{Equation 4.7}$$

As before, we set Equation 4.7 equal to zero and then solve for the equilibrium value of f:

$$ef(1-f) = p_i(1-f) \qquad \text{Expression 4.13}$$

$$ef = p_i \qquad \text{Expression 4.14}$$

Dividing both sides by e gives:

$$\hat{f} = \frac{p_i}{e} \qquad \text{Equation 4.8}$$

As in our original island–mainland model, persistence of the metapopulation is assured when there is both a propagule rain and a rescue effect. In fact, if the extinction parameter (e) is less than the probability of colonization (p_i), the metapopulation will be saturated at equilibrium, and all population sites will be occupied ($\hat{f} = 1$).

OTHER VARIATIONS

One final variation based on our simple metapopulation model would be to combine internal colonization with the rescue effect. In this case, the metapopulation is entirely closed to outside influences; both colonization and extinction probabilities are a function of the fraction of sites occupied. The equation for this model comes from substituting Expression 4.8 (internal colonization) and Expression 4.12 (rescue effect) into Equation 4.3 (the general model):

$$\frac{df}{dt} = if(1-f) - ef(1-f) \qquad \text{Equation 4.9}$$

However, if you try to set Equation 4.9 equal to zero and then solve for f, you will find there is no simple solution. Instead, the "equilibrium" depends on the relative sizes of i and e. If $i > e$, the immigration rate $[if(1-f)]$ will always be greater than the extinction rate $[ef(1-f)]$, so the metapopulation will "grow" until $f = 1$ (landscape saturation). Conversely, if $e > i$, the extinction rate exceeds the immigration rate, and the metapopulation will contract until $f = 0$ (regional extinction). If i and e vary stochastically, the metapopulation may fluctuate between these two equilibrium points (Hanski 1982). Finally, if i equals e, f will not change because the immigration rate will always equal the extinction rate. If some external force changes f, it will then stay at this new equilibrium value. We refer to this as a **neutral equilibrium**.

The metapopulation models that we have considered here have treated colonization as either internal or external. Similarly, extinctions were either independent or mediated by a rescue effect (Table 4.1). These four alternatives actually represent endpoints of a continuum. Colonization in most metapopulations is probably both from propagules generated from within the system and from propagules derived from external "mainland" sources. Similarly, there are extrinsic and intrinsic forces leading to extinction. These factors can be incorporated into a more general metapopulation model, which includes the four models developed in this chapter as special cases (Gotelli and Kelley 1993).

The derivations presented here just scratch the surface of metapopulation models (Hanski and Gilpin 1991). Other metapopulation models predict N directly, rather than just the presence or absence of populations. Metapopulation models have also been extended to two-species models of competitors or predator and prey. In some cases, species may coexist regionally that cannot coexist locally in closed populations. In other instances, exposing local populations to competitors or predators can lead to extinctions that might not have occurred otherwise. In Chapter 7, we will again return to a discussion of "open" systems when we model the colonization of an

Table 4.1 Four metapopulation models (Gotelli 1991).

| | | Extinction | |
		Independent	Mediated by rescue effect
Colonization	**External** ("propagule rain")	$\dfrac{df}{dt} = p_i(1-f) - p_e f$	$\dfrac{df}{dt} = p_i(1-f) - ef(1-f)$
	Internal	$\dfrac{df}{dt} = if(1-f) - p_e f$	$\dfrac{df}{dt} = if(1-f) - ef(1-f)$

island by an entire community of species. For now, we will return to our simple models of local populations and incorporate the effects of competitors (Chapter 5) and predators (Chapter 6).

Empirical Examples

THE CHECKERSPOT BUTTERFLY

Populations of the bay checkerspot butterfly (*Euphydryas editha bayensis*) occur in discrete patches that seem to be organized into a large metapopulation. The butterfly is somewhat of a habitat specialist; adult butterflies emerge in spring, and females prefer to lay their eggs on the annual plantain *Plantago erecta*. This host plant serves as a food source for the caterpillars, which feed for one or two weeks and then enter a summer diapause, or resting stage. Caterpillars resume feeding during the cool, rainy months of December to February, and then build cocoons. *P. erecta* grows in Northern California grasslands on serpentine soil rock outcroppings, which serve as potential population sites (Figure 4.3). Populations of the checkerspot butterfly have been studied in this area for over 30 years (Ehrlich et al. 1975).

Fluctuations in the weather can disrupt the life-cycle synchrony of the butterfly and its host plant, leading to local extinction. For example, at least three butterfly populations are known to have gone extinct following a severe drought in 1975–1977 (Murphy and Ehrlich 1980). Very small populations recorded in 1986 may represent successful recolonizations of empty sites (Harrison et al. 1988). The Morgan Hill site is a large patch of serpentine soil that supports a population of hundreds of thousands of butterflies. Because of its large size and the topographic diversity of the site, this population survived the drought and probably served as a source of colonists for empty patches.

The checkerspot metapopulation is similar, in some respects, to the island–mainland model, in which there is a persistent, external source of colonists. Although our simple metapopulation models assumed that all patches were identical, this was clearly not the case for the checkerspot butterfly. Populations were more likely to be found in sites that were close to the Morgan Hill population, had large areas of cool, north-facing slopes, and high densities of appropriate host plants (Harrison et al. 1988). For conservation purposes, preservation of the Morgan Hill population is probably essential because it provides colonists for other patches.

By their very nature, metapopulation studies require access to a lot of land. Although researchers have studied the checkerspot metapopulation for several years, work on many of the smaller patches can no longer be carried out. Attitudes of western land owners have changed; many are no longer willing to allow biologists onto their property to census the checkerspot butterfly

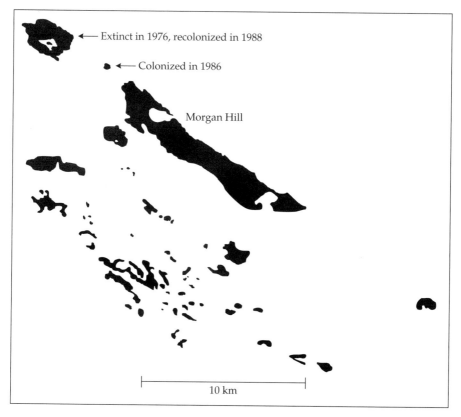

Figure 4.3 Distribution of serpentine soil grasslands in Santa Clara County, California. These habitat patches function as potential population sites for the bay checkerspot butterfly (*Euphydryas editha bayensis*). A large population of butterflies at the Morgan Hill site probably serves as a continual source of colonists for the other small patches, as in the simple island–mainland model. (After Harrison et al. 1988.)

(S. Harrison, personal communication). Some land owners fear that the discovery of an endangered species will deprive them of their property rights.

HEATHLAND CARABID BEETLES

Not all metapopulations occur in well-defined patches, as in the checkerspot butterfly example. Populations may be organized as a metapopulation even in the absence of discrete habitat patches. In the northern Netherlands, populations of carabid ground beetles have been studied by pitfall trapping for over 35 years (den Boer 1981). Radioactive marking revealed that most individuals moved a very limited distance. For example, 90% of the individuals of the beetle *Pterostichus versicolor* moved less than 100 meters a day.

Consequently, sites separated by even modest distances effectively contain different subpopulations that are connected by migration.

Figure 4.4a shows the size of 19 subpopulations of *P. versicolor* that were studied for 21 years. Although populations fluctuated asynchronously, there

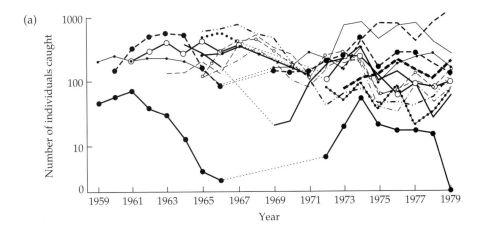

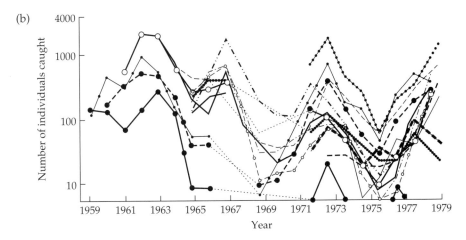

Figure 4.4 (a) Metapopulation dynamics of the ground beetle *Pterostichus versicolor* in heathlands of the northern Netherlands. Each symbol represents the track for a different subpopulation in the heath. Note the great variability in population dynamics and the relative rarity of local extinctions. Broken lines indicate gaps in sampling. Lines that touch the *x* axis indicate local extinctions. At each time period, some subpopulations are usually increasing, which may act as sources of migrants that prevent the extinction of declining subpopulations. (b) Metapopulation dynamics of the ground beetle *Calathus melanocephalus* in heathlands of the northern Netherlands. In contrast to *P. versicolor*, subpopulations of *C. melanocephalus* tend to fluctuate in synchrony. Consequently, there are no "source" areas to rescue declining subpopulations, so that local extinctions are more frequent. (After den Boer 1981.)

were almost no extinctions recorded during this time period. This is because, at any point in time, some populations were increasing in size and acting as **source populations** that prevented the extinction of other, declining **sink populations**. In contrast, the population fluctuations of the species *Calathus melanocephalus* were much more synchronous during this period (Figure 4.4b). As a consequence, conditions were sometimes uniformly bad for all populations. At these times, there were no source populations available, so population extinctions were much more frequent. Because each subpopulation of *C. melanocephalus* behaved similarly, the risk of extinction was high. Because each subpopulation of *P. versicolor* behaved differently, the metapopulation structure effectively spread the risk of extinction. We still don't understand why the population dynamics of these two beetle species are so different, but it is clear that metapopulation structure affects local extinction and perhaps long-term persistence.

Problems

4.1. You are studying a rare and beautiful species of ant lion (see cover). Populations of the ant lion live on a set of islands and on an adjacent mainland that serves as a permanent source of colonists. You can assume that the mainland is the only source of colonists and that extinctions on the islands are independent of one another.

   a. If $p_i = 0.2$ and $p_e = 0.4$, calculate the fraction of islands occupied at equilibrium.

  *b. A developer is preparing to pave over the mainland area for a new condominium complex. To appease local environmental groups, the developer promises to set aside the islands as a permanent "ant lion nature reserve." Assuming that $p_e = 0.4$ and $i = 0.2$, predict the fate of the island populations after the mainland population is eliminated.

4.2. An endangered population of 100 frogs lives in a single pond. One proposal for conserving the frog population is to split it into three populations of 33 frogs, each in a separate pond. You know from your demographic studies that decreasing the frog population from 100 to 33 individuals will increase the yearly risk of extinction from 10% to 50%. In the short run, is it a better strategy to retain the single population or to split it into three?

*4.3. Suppose a metapopulation has a propagule rain and a rescue effect. The parameters are $p_i = 0.3$ and $e = 0.5$. Forty percent of the population sites are occupied. Is this metapopulation expanding or shrinking?

* Advanced problem

CHAPTER 5

Competition

Model Presentation and Predictions

COMPETITIVE INTERACTIONS

Chapters 1 through 4 examined single-species population growth. Although we didn't exclude the possibility that other species were important, we did not write explicit equations for populations of predators, prey, or competitors. Instead, the effects of other species were contained in constants such as K, the carrying capacity of the environment (Chapter 2), or p_e, the probability of local population extinction (Chapter 4). In this chapter, we will introduce a second population of a competing species and model the growth of two interacting populations.

Before introducing the model, we need to specify exactly what we mean by "competition." **Competitive interactions** are those in which two species negatively influence each other's population growth rates and depress each other's population sizes. This general definition encompasses a variety of population interactions. **Exploitation competition** occurs when populations depress one another through use of a shared resource, such as food or nutrients. Examples include tropical reef fish that graze on the same kinds of algae, and desert plants that compete for a limited supply of water.

Interference competition occurs when an individual or population behaves in a way that reduces the exploitation efficiency of another individual or population. Examples include song birds that maintain well-established breeding territories, and ant colonies that kill invaders at food patches. Even plants engage in a form of interference competition known as **allelopathy**. Many plant species, particularly aromatic herbs, release toxic chemicals that poison the soil for competitors. The key element in interference competition is that species supress one another directly, not only through their indirect use of resources.

Interference competition leaves more resources for the winner to consume, so it may evolve as an adaptation when exploitation competition is severe. As an analogy for understanding these two kinds of interactions, exploitation competition is when you and a friend are sitting at a table drinking the same milkshake with straws. The "winner" in exploitation competition is the one who consumes the most milkshake. Interference competition is when you reach over and pinch your friend's straw!

Pre-emptive competition is a third category that has elements of both exploitation and interference. In pre-emptive competition, organisms compete for space as a limiting resource. Examples include birds that use tree holes for nesting and intertidal algae that must attach to stable rock surfaces. Unlike food or nutrients that are used exploitatively, space is a renewable resource that is "recycled"—as soon as an organism dies or leaves, the space is immediately available for use by other individuals.

We need to consider not only the mechanism of competition, but the extent to which competition occurs within and between species. **Intraspecific competition** is competition that occurs among members of the same species. The logistic equation (Equation 2.1) is a model of intraspecific competition because the per capita growth rate diminishes as the population becomes more crowded. **Interspecific competition** is competition between individuals of two or more different species. In this chapter we will build a model of interspecific competition that is a direct extension of the logistic equation.

THE LOTKA–VOLTERRA COMPETITION MODEL

In the 1920s and 1930s, Alfred J. Lotka (1880–1949) and Vito Volterra (1860–1940) described a simple mathematical model of interspecific competition that is the framework for competition studies in ecology. The model treats populations of two competing species, which we will designate as N_1 and N_2. Each population grows according to the logistic, with its own intrinsic rate of increase (r_1 or r_2) and its own carrying capacity (K_1 or K_2). As in the logistic model, population growth is reduced by intraspecific competition:

$$\frac{dN_1}{dt} = r_1 N_1 \left(\frac{K_1 - N_1}{K_1} \right) \qquad \text{Expression 5.1}$$

$$\frac{dN_2}{dt} = r_2 N_2 \left(\frac{K_2 - N_2}{K_2} \right) \qquad \text{Expression 5.2}$$

In our new model, the population growth rate is further depressed by the presence of the second species. For now, assume that the growth is reduced by some function (f) of the number of individuals of the competitor:

$$\frac{dN_1}{dt} = r_1 N_1 \left(\frac{K_1 - N_1 - f(N_2)}{K_1} \right) \qquad \text{Expression 5.3}$$

$$\frac{dN_2}{dt} = r_2 N_2 \left(\frac{K_2 - N_2 - f(N_1)}{K_2} \right) \qquad \text{Expression 5.4}$$

These expressions show that population growth rate is depressed by both intraspecific and interspecific competition. There are many complicated functions that we could use in Expressions 5.3 and 5.4, but the simplest formula is to multiply the population size of the competitor by a constant number:

$$\frac{dN_1}{dt} = r_1 N_1 \left(\frac{K_1 - N_1 - \alpha N_2}{K_1} \right) \qquad \text{Equation 5.1}$$

$$\frac{dN_2}{dt} = r_2 N_2 \left(\frac{K_2 - N_2 - \beta N_1}{K_2} \right)$$

Equation 5.2

COMPETITION COEFFICIENTS

The **competition coefficients** α and β are critical to understanding the Lotka–Volterra model. α is a measure of the effect of species 2 on the growth of species 1. If $\alpha = 1$, then individuals of the two species are interchangable—each has an equal effect in depressing the growth of species 1. On the other hand, suppose that $\alpha = 4$. Each individual of species 2 that is added to the environment depresses the growth of N_1 by the same amount as adding four individuals of species 1. Thus, α is a measure of the relative importance *per individual* of interspecific and intraspecific competition. If $\alpha > 1$, the per capita effect of interspecific competition is greater than the per capita effect of intraspecific competition. If $\alpha < 1$, the intraspecific competition is more important—population growth of species 1 is depressed more by the addition of an individual of N_1 than by addition of an individual of the competing species. Finally, notice that if $\alpha = 0$, there is no competitive effect, and Equation 5.1 reduces to the single-species logistic equation (Equation 2.1). Thus, we can define α as the per capita effect of species 2 on the population growth of species 1, *measured relative to the effect of species 1.*

Similar reasoning applies to the interpretation of β, which is the per capita effect of species 1 on the population growth of species 2. It is important to realize that α and β need not have the same values. Competitive effects in nature often are asymmetrical—adding an individual of one species may severely depress the population growth of a second species, whereas the reverse is not true. Although both species in our model coexist in the same location, remember that they each have separate carrying capacities (K_1 and K_2), and intrinsic rates of increase (r_1 and r_2). Although r_1 and r_2 do not affect the outcome of competition in this model, we will see in the next section that the carrying capacities and competition coefficients are critical for determining species coexistence.

A useful way to understand α and β is to return to the analogy we developed in Chapter 2 (Krebs 1985): the carrying capacity of the environment for species 1 is a square frame that holds a limited number of flat tiles (individuals). In our competition model, the tiles come in two different sizes, representing the two different species (Figure 5.1). Continuing the analogy, α is the relative area of the two tiles. For example, if $\alpha = 4$, then a single individual of species 2 consumes four times the remaining resources of the environment as a single individual of species 1. So, a tile of species 2 has four times the area of a tile of species 1. At equilibrium, the frame is filled with a mix of the

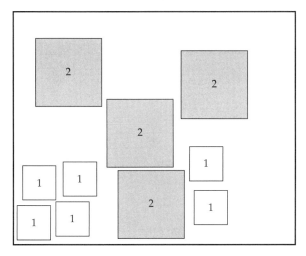

Figure 5.1 A graphical analogy for interspecific competition. The heavy square frame represents the carrying capacity for species 1 (K_1). Each individual consumes a portion of the limited resources available and is represented by a tile. Individuals of species 2 reduce the carrying capacity four times as much as individuals of species 1. Hence, the tiles for species 2 are four times larger than those for species 1, and $\alpha = 4.0$. (After Krebs 1985.)

two kinds of tile. In the next section we will solve for these equilibrium densities.

EQUILIBRIUM SOLUTIONS

As in all our previous analyses, we find the equilibrium population densities ($\hat{N}$) by setting the differential equations equal to zero and solving for N:

$$\hat{N}_1 = K_1 - \alpha N_2 \qquad \text{Equation 5.3}$$

$$\hat{N}_2 = K_2 - \beta N_1 \qquad \text{Equation 5.4}$$

These results make intuitive sense. The equilibrium for N_1 is the carrying capacity for species 1 (K_1) reduced by some amount due to the presence of species 2 (αN_2). But we have trouble putting numbers into these solutions—the equilibrium for species 1 depends on the equilibrium for species 2, and vice versa! We can make progress by substituting the equilibrium for N_2 into Equation 5.3, so that the answer will be entirely in terms of N_1:

$$\hat{N}_1 = K_1 - \alpha(K_2 - \beta \hat{N}_1) \qquad \text{Expression 5.5}$$

Similarly, we can substitute the equilibrium for N_1 into Equation 5.4:

$$\hat{N}_2 = K_2 - \beta(K_1 - \alpha\hat{N}_2) \qquad \text{Expression 5.6}$$

For each of these expressions, we carry out the multiplication, move all the N terms to the left side of each equation, and arrive at the following solutions:

$$\hat{N}_1 = \frac{K_1 - \alpha K_2}{1 - \alpha\beta} \qquad \text{Equation 5.5}$$

$$\hat{N}_2 = \frac{K_2 - \beta K_1}{1 - \alpha\beta} \qquad \text{Equation 5.6}$$

Note that for both species to have an equilibrium population size greater than zero, the denominator of each expression must usually be greater than zero. Thus, it is usually the case that the product $\alpha\beta$ must be less than 1 for both species to coexist.

THE STATE SPACE

Although Equations 5.5 and 5.6 tell us the equilibrium conditions for the Lotka–Volterra competition models, they do not provide much insight into the dynamics of competitive interactions, or whether these equilibrium points are stable or not.

We can understand these equations much better by using the **state-space graph**, a special kind of plot. In the state-space graph, the x axis represents the abundance of species 1, and the y axis represents the abundance of species 2. This graph takes a bit of getting used to, but it is an important tool in multi-species models. We will use it again in Chapter 6, when we explore predator–prey models.

What do points in state space represent? A point in this graph represents a *combination of abundances* of species 1 and species 2. The abundance of species 1 can be read from the x axis and the abundance of species 2 can be read from the y axis. If our point falls on the x axis, then only species 1 is present and the abundance of species 2 is zero. For points on the y axis, only species 2 is present. So, the full collection of points in this graph represents all the different combinations of species 1 and species 2 that we could put together.

We use the state-space graph to understand the population dynamics of two competitors. Imagine two competing species whose populations are changing size with time. At each point in time, we could represent their abundances by a single point in the state space (Figure 5.2a). As both populations change in size (Figure 5.2b), we would trace a line in the state space. The final

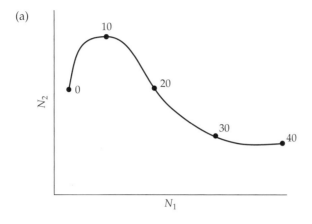

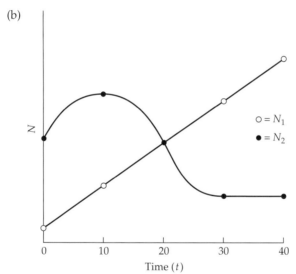

Figure 5.2 (a) A state-space graph. The axes of the state space are the abundances of the two species (N_1 and N_2). As abundances change through time, a curve is traced from left to right. The numbers on the curve indicate time, beginning at 0 and ending at 40. (b) Translation of the state-space graph in (a). The abundances of each species are read from the state-space graph at different times. Note that species 2 first increases and then decreases, whereas species 1 shows a continuous increase in population size.

equilibrium point is the end of this line, and if either species goes extinct, this point falls on one of the two axes of the state-space graph.

How can we use the state-space graph to help us understand the Lotka–Volterra equations? We will first plot Equation 5.3 in the state space. Equation 5.3 is the equilibrium solution for species 1, and its graph is a straight line.

This line represents the combinations of abundances of species 1 and species 2 for which there is zero growth of species 1. At any point on this line, the carrying capacity for species 1 is entirely "filled" with individuals of both species. This line is an **isocline**: a set of abundances for which the growth rate (dN/dt) of one species is zero.

The isocline for species 1 intersects the axes of the state-space graph in two places. The intersection on the x axis is at a value of K_1. This equilibrium point represents the case in which species 2 is absent and species 1 has grown to its own carrying capacity. The other point is the intersection on the y axis. Here, species 1 is essentially extinct, and the carrying capacity of species 1 is entirely filled with individuals of species 2. The equilibrium solution at this point is K_1/α individuals of species 2 and zero individuals of species 1. Between these extremes are combinations of both species that fall on the isocline (Figure 5.3).

The isocline for species 1 splits the state space into two regions. If we are to the left of the isocline, the joint abundance of N_1 and N_2 is less than the car-

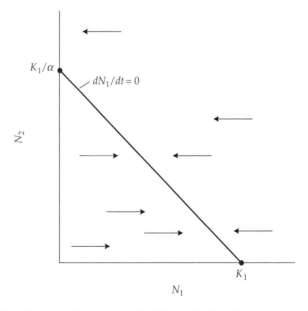

Figure 5.3 The linear isocline for species 1 in the Lotka–Volterra competition model. The isocline defines the combination of abundances for which species 1 shows zero growth. For points to the left of this line, the population of species 1 increases, indicated by the right-pointing horizontal arrow. For points to the right, the joint abundance of species 1 and species 2 exceeds the isocline for species 1, so its population decreases, indicated by the left-pointing arrows.

rying capacity for species 1, so N_1 will increase. An increase in N_1 in the state space is represented as a *horizontal* arrow pointing to the right. The arrow is horizontal because the abundance of species 1 is represented on the x axis. When you work with state-space graphs, pay close attention to which species' isocline you are plotting. Any point to the left of the isocline for species 1 generates a horizontal right-pointing arrow. Under these circumstances, we know that species 1 has a positive growth rate, so its population will increase in size. In contrast, if we are to the right of the isocline, the joint abundance of N_1 and N_2 exceeds the carrying capacity of species 1. In this case, the growth rate of N_1 is negative, and the population decreases. The decrease is represented as a left-pointing horizontal arrow in the state space. Finally, if we are at a point precisely on the isocline, N_1 neither increases nor decreases, and there is no movement in the horizontal direction.

Now we plot the isocline for species 2 in the state space. The isocline of species 2 intersects the y axis at a value of K_2 and intersects the x axis at a value of K_2/β. The first case is one in which species 1 is absent and species 2 has grown to its carrying capacity. In the second case, species 2 is absent, and its carrying capacity is occupied by K_2/β individuals of species 1. Once again, the isocline for species 2 splits the state space into two regions. If we are below the isocline, the joint abundance of species 1 and species 2 is below K_2, and N_2 will increase. Because species 2 is on the y axis, positive growth of species 2 is represented as a *vertical* arrow pointing up in state space. Similarly, if we are above the isocline, the carrying capacity of species 2 is exceeded; its population decreases, represented by a downward-pointing arrow (Figure 5.4).

It is important to recognize that there is a unique isocline for each species that dictates its population growth. By plotting both isoclines together in the state space, we can understand the dynamics of two-species competition. Of course, there are an infinite number of isoclines we could build, simply by using different values of K_1, K_2, α, and β. Fortunately, there are only four qualitatively different ways we can plot the isoclines. These four patterns represent the four possible outcomes of competition in the Lotka–Volterra equations.

GRAPHICAL SOLUTIONS TO THE LOTKA–VOLTERRA COMPETITION MODEL

Case 1: Species 1 wins in competition. Figure 5.5 shows one possible configuration of the two isoclines in the state space: the isocline for species 1 lies entirely above the isocline for species 2. In this case, the state space is split into three regions. If we are in the lower left-hand region of the graph, we are below the isoclines of both species, and both species can increase. This is represented by a horizontal and vertical arrow joined at their base. The joint movement of these two populations is represented by the vector sum, which is an arrow that points towards the upper right-hand corner of the graph.

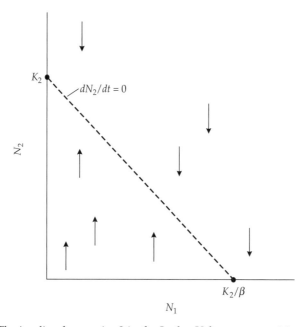

Figure 5.4 The isocline for species 2 in the Lotka–Volterra competition model. Note that the arrows point vertically for species 2, because its abundance is measured on the y axis of the state space graph.

Conversely, if we are in the upper right-hand region of the state space, we are above the isoclines of both species. Both populations will decrease, and the joint vector points towards the origin of the graph.

Things get more interesting in the interior region. Here, we are *below* the isocline of species 1, so its population increases in size, and the horizontal arrow points to the right. However, we are *above* the isocline of species 2, so its population decreases, and the vertical arrow points down. The joint vector points down and to the right, which takes the populations towards the carrying capacity of species 1. Eventually, species 2 declines to extinction, and species 1 increases to K_1. Notice that, no matter what combination of abundances we start with, the arrows always point towards this outcome. If the isocline of species 1 lies above that of species 2, species 1 always wins in competition, and species 2 is driven to extinction.

Case 2: Species 2 wins in competition.   If we graph the isocline of species 2 above that of species 1, then we reverse the conditions and species 2 wins in competition (Figure 5.6). The only difference in this graph is the vector in the interior region. In this case, we are *above* the isocline of species 1, which generates a horizontal arrow to the left, but we are *below* the isocline of species 2, which gener-

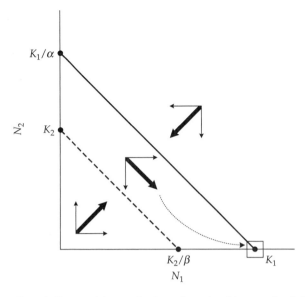

Figure 5.5 Case 1: Competitive exclusion of species 2 by species 1. The thin arrows show the trajectories of each population, and the thick arrow is the joint vector of movement. Competition results in the exclusion of species 2 and an equilibrium for species 1 at carrying capacity. The box indicates a stable equilibrium point.

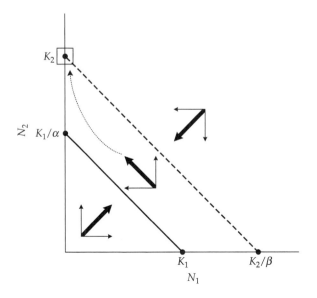

Figure 5.6 Case 2: Competitive exclusion of species 1 by species 2.

ates a vertical arrow pointing up. The joint vector points up and to the left, taking us towards the equilibrium point at K_2, with N_1 going extinct.

Case 3: Coexistence in a stable equilibrium. The remaining two cases are slightly more complex, because they involve isoclines that cross, dividing the state space into four regions. Nevertheless, the analysis is exactly the same. We simply plot the vectors in each of the four regions to determine the outcome (Figure 5.7). First, note that because the two isoclines cross, there must be an equilibrium point—the crossing of the isoclines represents a combination of abundances for which both species 1 *and* species 2 have achieved zero growth. The state space analysis reveals whether that equilibrium is stable or not.

As in our previous two examples, the region close to the origin is one of joint growth of both populations, and the region in the upper right-hand corner of the graph is one of joint decrease. The vectors in these regions point towards the equilibrium intersection. If we are in the region of the graph on the lower right, we are *above* the isocline of species 1, but *below* the isocline of species 2. Here, the joint vector points towards the center, as N_1 decreases along the horizontal axis and N_2 increases along the vertical axis. Finally, if

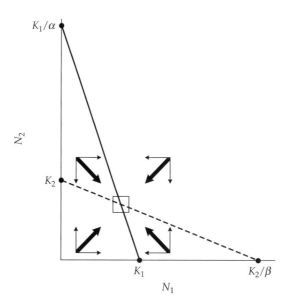

Figure 5.7 Case 3: Coexistence in a stable equilibrium. The two isoclines cross, and the joint vectors point in towards the equilibrium point. The equilibrium is stable because if the populations are displaced, they will always return to their equilibrium sizes.

we are in the region of the graph on the upper left, we are *above* the isocline of species 2, but *below* the isocline of species 1, and the joint vector again points towards the center.

This is a **stable equilibrium** in which all roads lead to Rome—no matter what the initial abundances of the two species are, both populations will move towards the joint equilibrium value. Although this equilibrium is stable and both species coexist, note that each species persists at a lower abundance than it would in the absence of its competitor. Competition reduces the population size of each species, but neither can drive the other extinct.

Case 4: Competitive exclusion in an unstable equilibrium. This final case is the one in which the isoclines cross in the opposite way (Figure 5.8). Once again, both populations increase in the sector closest to the origin, and both populations decrease in the upper right-hand region. But the pattern changes in the two remaining slivers of state space. In the lower right-hand region, we are *below* the isocline of species 1, but *above* the isocline of species 2. In this region of the graph, the populations move *away* from the joint equilibrium and towards K_1. Similarly, in the fourth region of the state space, we are *above* the isocline for N_1, but *below* that for N_2. The populations move away from the joint equilibrium and towards K_2.

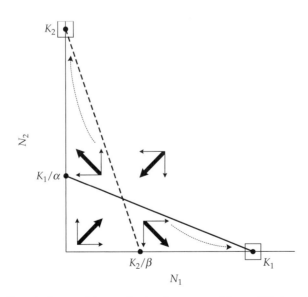

Figure 5.8 Case 4: Competitive exclusion in an unstable equilibrium. The two isoclines again cross and form an equilibrium point. However, the joint vectors point away from this equilibrium. If the populations are displaced, one species or the other will win in competition, depending on the starting abundances.

Case 4 represents an **unstable equilibrium**. If the populations are displaced from the joint equilibrium, they will eventually end up in one of the two regions of the graph that will take them to competitive exclusion. Thus, both species cannot persist in the long run, and one will be driven to extinction by competition. However, the winner is difficult to predict. The population that has a numerical advantage is the one that will probably win in competition, but the outcome depends on the initial position in the state space, and the relative growth rates of the two competitors (r_1 and r_2).

THE PRINCIPLE OF COMPETITIVE EXCLUSION

Now that we understand the four graphical solutions to the Lotka–Volterra competition equations, we will take another look at the algebraic solutions. We can reason that species 1 will always persist if it can invade under the worst possible circumstances. The worst scenario for species 1 is that its own abundance is close to zero ($N_1 \approx 0$), and the abundance of its competitor is close to carrying capacity ($N_2 \approx K_2$). If N_1 can achieve a positive per capita growth rate $[(dN_1/dt)(1/N_1) > 0]$ under these circumstances, then it should always be able to invade (MacArthur 1972). Plugging these conditions into Equation 5.1 gives:

$$\left(\frac{dN_1}{dt}\right)\left(\frac{1}{N_1}\right) = r_1\left(\frac{K_1 - 0 - \alpha K_2}{K_1}\right) \qquad \text{Expression 5.7}$$

Since r_1 is always positive, the following inequality must hold for N_1 to increase:

$$\frac{K_1 - \alpha K_2}{K_1} > 0 \qquad \text{Expression 5.8}$$

which reduces to:

$$\frac{K_1}{K_2} > \alpha \qquad \text{Expression 5.9}$$

If species 1 is to successfully invade, the ratio of the carrying capacities must exceed the competitive effect of species 2 on species 1. In other words, if species 2 is a strong competitor, species 1 must have a relatively large carrying capacity to persist.

Using Equation 5.2, we can go through a similar calculation to arrive at the following inequality for the persistence of species 2:

$$\frac{K_2}{K_1} > \beta \qquad \text{Expression 5.10}$$

Flipping the inequality makes this directly comparable with Expression 5.9:

$$\frac{1}{\beta} > \frac{K_1}{K_2} \qquad \text{Expression 5.11}$$

Table 5.1 Algebraic inequalities defining the ability of species to invade and the outcome of competition in the Lotka–Volterra equations.

(a)

Inequality	Outcome
$\dfrac{K_1}{K_2} > \alpha$	Species 1 invades
$\dfrac{K_1}{K_2} < \alpha$	Species 1 cannot invade
$\dfrac{K_1}{K_2} < \dfrac{1}{\beta}$	Species 2 invades
$\dfrac{K_1}{K_2} > \dfrac{1}{\beta}$	Species 2 cannot invade

(b)

Species 1 invades	Species 2 invades	Inequality	Outcome
Yes	No	$\dfrac{1}{\beta} < \dfrac{K_1}{K_2} > \alpha$	Species 1 wins (Case 1)
No	Yes	$\dfrac{1}{\beta} > \dfrac{K_1}{K_2} < \alpha$	Species 2 wins (Case 2)
Yes	Yes	$\dfrac{1}{\beta} > \dfrac{K_1}{K_2} > \alpha$	Stable coexistence (Case 3)
No	No	$\dfrac{1}{\beta} < \dfrac{K_1}{K_2} < \alpha$	Unstable equilibrium (Case 4)

Now we have expressions for whether N_1 will invade or not, and whether N_2 will invade or not. Putting these expressions together generates four algebraic inequalities that define the four graphical solutions to the Lotka–Volterra equations. For example, if species 1 can invade ($K_1/K_2 > \alpha$), but species 2 cannot ($1/\beta < K_1/K_2$), then we have defined the conditions for case 1, in which species 1 always wins in competition. If both species are able to invade, we have the stable coexistence of case 3, whereas if neither species can invade, we have the unstable equilibrium of case 4 (Table 5.1).

These inequalities give us insight into one of ecology's enduring proverbs, the **principle of competitive exclusion**. Briefly stated, the principle is that "complete competitors cannot coexist" (Hardin 1960). In other words, if species are able to coexist, there must be some difference between them in resource use (Gause 1934).

If two species are very similar in their resource use, then α and β should be very close to 1. Suppose, for example, that $\alpha = \beta = 0.9$. From the inequality in Table 5.1, coexistence of these species requires that:

$$\frac{1}{0.9} > \frac{K_1}{K_2} > 0.9$$

Expression 5.12

$$1.1 > \frac{K_1}{K_2} > 0.9$$

Expression 5.13

Thus, if the species are very similar in their use of resources, there is only a narrow range of carrying capacities that will ensure stable coexistence. In contrast, suppose that $\alpha = \beta = 0.2$, indicating that species differ greatly in their use of common resources. In this case, coexistence will occur if:

$$5 > \frac{K_1}{K_2} > 0.2$$

Expression 5.14

In this case, the two species will coexist with a wide range of possible carrying capacities. Thus, our analysis of the Lotka–Volterra equations allows us to refine the competitive exclusion principle: the more similar species are in their use of shared resources, the more precarious their coexistence.

The Lotka–Volterra equations are the simplest two-species model of competition. As you might expect, it is even more difficult to obtain coexistence of species in models that have three or more competitors. For many years, ecologists have studied the "coexistence problem," and discovered that species often coexist in nature with little apparent difference in resource exploitation. In these circumstances, one or more of the following assumptions of the model has been violated.

Model Assumptions

As in the logistic and exponential growth models, we assume there is no age or genetic structure to the populations, no migration, and no time lags. The following assumptions also apply to the Lotka–Volterra model:

✔ **Resources are in limited supply.** The result of resource limitation is both intra- and interspecific competition. If resources are not limiting, then an infinite number of species can coexist, regardless of how similar they are in resource use.

✔ **Competition coefficients (α and β) and carrying capacities (K_1 and K_2) are constants.** If these parameters should change with time or density, it may be difficult to predict species coexistence.

✔ **Density dependence is linear.** Adding an individual of either species produces a strictly linear decrease in per capita population growth rate. This is reflected in the linear isoclines of the Lotka–Volterra model. Models with nonlinear isoclines have more complex stability properties that are not easy to deduce from simple state-space graphs.

Model Variations

INTRAGUILD PREDATION

Ecologists classify species interactions according to their effects on population growth rate. Thus, competition is defined as both species having a net negative effect on one another (–,–), mutualism as both species having a net positive effect (+,+), and predation or parasitism as one species gaining and the other species losing (+,–). These classifications are convenient and natural, and they reflect our model assumptions that interaction coefficients are constant and that there is no age structure in the populations.

But when we study the natural history of many animals, we find they cannot be classified simply as "predators" or "competitors." For example, lions prey on the young of cheetahs, wild dogs, and spotted hyenas, but also compete with these same species for prey. Flour beetles in the genus *Tribolium* compete for food, but at high densities they also consume one another's larvae. For many predators, diet is determined strictly by their size and what they can get their jaws around. As individuals age, their diets can change radically. Anyone who has tried to raise baby fish in an aquarium can appreciate that predation is often critically tied to body size. Individuals of a single species may act as prey, competitors, or predators, depending on their age and size. **Intraguild predation** (IGP) is the ecological interaction in which two competing species also interact as predator and prey. IGP is not an isolated phenomenon; it is common in terrestrial, marine, and freshwater communities, and probably represents the rule rather than the exception in nature (Polis et al. 1989).

How can we modify our simple competition model to take account of IGP? Suppose that two species compete according to the Lotka–Volterra equations, but species 1 is also a predator on species 2. This is a simple model that does not involve age structure, reciprocal predation, or cannibalism. However, it at least illustrates the way that IGP can modify ecological interactions. The growth equation for species 1 ("predator") is:

$$\frac{dN_1}{dt} = r_1 N_1 \left(\frac{K_1 - N_1 - \alpha N_2}{K_1} \right) + \gamma N_1 N_2 \qquad \text{Equation 5.7}$$

This is identical to the original Lotka–Volterra model, except we have added an additional term. This addition represents the increase in growth rate that species 1 receives by feeding on species 2. The amount of this increase depends on the abundances of predator and prey ($N_1 N_2$) and an interaction coefficient (γ). We will see a similar expression in Chapter 6 when we build a predator-prey model. The growth equation for species 2 ("prey") is:

$$\frac{dN_2}{dt} = r_2 N_2 \left(\frac{K_2 - N_2 - \beta N_1}{K_2} \right) - \delta N_1 N_2 \qquad \text{Equation 5.8}$$

Again, growth of species 2 is described by the Lotka–Volterra model, but is further reduced because of losses due to predation by species 1. These losses also depend on the abundances of predator and prey ($N_1 N_2$) and an interaction coefficient (δ). Note that the interaction coefficients for predator (γ) and prey (δ) need not be equivalent. The loss of an individual to predation usually does not correspond to a symmetrical gain for the predator population. Again, these ideas are explored more thoroughly in Chapter 6.

How does IGP affect the coexistence of species? The graphical effect of IGP is to rotate the isoclines. IGP does not change the carrying capacity for either predator or prey. Instead, it changes the abundance of the competitor that is necessary to cause extinction. Consequently, each isocline is rotated up or down, but remains fixed at the intercept on its own axis. For the predator, the isocline swings up, because it now requires more competitors to drive the predator to extinction than before (Figure 5.9a). For the prey species, IGP swings the isocline in towards the origin, because it now requires fewer competitors to cause extinction (Figure 5.9b).

IGP can either reinforce or reverse the outcome of competition, depending on the position of the isoclines and the amount of rotation (which is ultimately controlled by the interaction coefficients). For example, if the inferior competitor is also the prey species, IGP merely adds the insult of predation to the injury of competition and reinforces the extinction of species 2 (Figure 5.10a). But if the inferior competitor is the predator, IGP can change the outcome from competitive exclusion (case 1) to stable coexistence (case 3; Figure 5.10b). Other outcomes are possible, and IGP may provide insight into species coexistence when simple competition and predation models fail (Polis et al. 1989).

Empirical Examples

COMPETITION BETWEEN INTERTIDAL SANDFLAT WORMS

In northern Puget Sound, many species of marine worms coexist in intertidal sandflats at very high densities. Abundances can be manipulated experi-

(a)

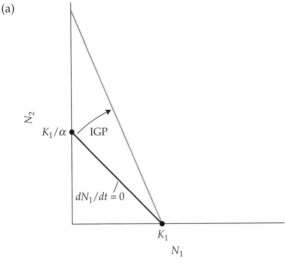

(b)

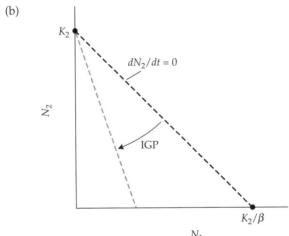

Figure 5.9 (a) Intraguild predation rotates the isocline of the "predator" species up, because it now requires more individuals of the competitor-prey to drive it to extinction. (b) Intraguild predation rotates the isocline of the "prey" species down, because it now requires fewer individuals of the competitor-predator to drive it to extinction.

mentally, allowing for a direct test of the Lotka–Volterra competition model. Gallagher et al. (1990) examined competition between juveniles of the poly-chaete *Hobsonia florida* and a number of closely related species of oligochaetes. Both *Hobsonia* and the oligochaetes coexist in dense aggregations in nature and feed on benthic diatoms.

Gallagher et al. (1990) used field experiments to determine whether such coexistence could be successfully predicted by the Lotka–Volterra model. By adding predatory shrimp to small (26-centimeter diameter) field enclosures, the authors were able to manipulate the densities of *Hobsonia* and the

(a)

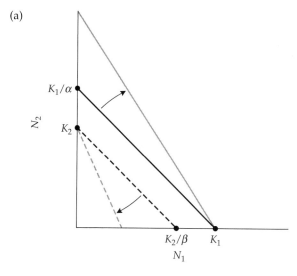

(b)

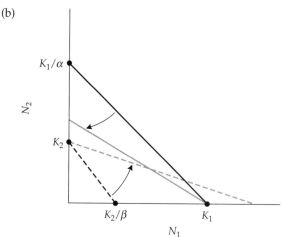

Figure 5.10 (a) Intraguild predation reinforces competitive exclusion. In this example, the superior competitor (N_1) is also the predator, so the shifted isoclines lead to the same outcome. (b) Intraguild competition reverses competitive exclusion. In this example, the inferior competitor (N_2) is now the predator. The isoclines shift from competitive exclusion (Case 1) to stable coexistence (Case 3).

oligochaetes in the patch. These starting densities represented a single point in the state space. Next, they measured the increase and decrease of each population in the patch after three days. These changes revealed the vector of population dynamics in the state space. By repeating this procedure for different starting densities, they were able to determine the placement of both isoclines. These field experiments produced the following estimates: K_1 (*Hobsonia*) = 64.2, α = 1.408; K_2 (oligochaetes) = 50.7, β = 0.754. Finally, the authors started two patches at a low initial abundance of both competitors and then followed their dynamics for 55 days.

The isoclines for both species are plotted in Figure 5.11. Superimposed on this state space is one of the trajectories for the 55-day experiment. Because the isocline for the oligochaetes lies slightly above the isocline for *Hobsonia*, the model predicts that the oligochaetes should win in competition. But the trajectory of the 55-day experiment did not reach the oligochaete carrying capacity, and in nature, both species coexist. The simple Lotka–Volterra model must be rejected for this system.

Why did the model fail to give us the correct predictions? Because the isoclines of the two species are very close to one another, the predicted time to

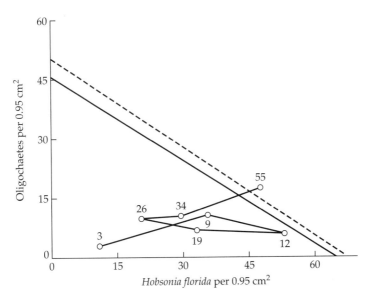

Figure 5.11 Competition between marine intertidal worms. The solid line is the estimated isocline for *Hobsonia florida*, and the dashed line is the estimated isocline for the oligochaetes. The line segments trace an experiment in the state space that was started with low abundances of both competitors. The numbers indicate the number of days since the start of the experiment. (From Gallagher et al. 1990.)

extinction is long. Moreover, there are seasonal changes in diatom abundance, so that the carrying capacities for each species are always changing. When carrying capacities change, the isoclines "wobble" through time, so that population trajectories may be continually changing. Under these conditions, there may not be enough time for one species to win in competition. Thus, the oligochaetes do not competitively exclude *Hobsonia* because the environment is always changing. As the ecologist G. Evelyn Hutchinson (1967) wrote: "The competitors of a given genus or other higher taxon are from time to time lined up, and sometimes the race begins, but as it might be in the works of Lewis Carroll, the event is always called off before it is completed and something entirely different is arranged in its place."

THE SHAPE OF A GERBIL ISOCLINE

Gerbils are mouselike rodents of the deserts of Africa and the Middle East. They are nocturnal seed foragers, and the coexistence of several gerbil species may depend on their use of common food and habitat resources. Abramsky et al. (1991) studied the coexistence of *Gerbillus allenbyi* and *G. pyramidum* in the western Negev desert of Israel.

Experimental studies of vertebrate competition are particularly difficult because of the large areas needed to enclose populations, and because competition is often mediated by subtle behavioral interactions. Abramsky et al. (1991) took advantage of the fact that *G. pyramidum* is considerably larger than *G. allenbyi* (mean mass = 40 grams versus 26 grams). The authors built enclosures that were 100 meters on a side (one hectare in area). Each enclosure was separated into two plots by a common fence. This fence had small gates to permit gerbils to move between the two sides. The gates were large enough to allow *G. allenbyi* through, but too small for *G. pyramidum* to pass. Thus, the fence acted as a semipermeable membrane, allowing *G. allenbyi* to "equilibrate" its density on the two sides based on the density of *G. pyramidum*.

Although the Lotka–Volterra competition model predicts changes in population growth rate, these are difficult to measure in short-term experiments on vertebrates. Moreover, the effects of competition on gerbil populations are likely to be expressed more immediately in changes in behavior and foraging activity. Instead of measuring gerbil density, the authors measured the "activity density" of each species by counting gerbil footprints in clean trays of sand that were placed in the plots each night. This index was correlated with density and foraging activity of individual gerbils.

The authors established one half of each enclosure with a high density of *G. pyramidum* and the other half with a low density. The density of *G. allenbyi* was allowed to equilibrate to these differences in competitor density. The resulting changes in activity of both species can be plotted in state space.

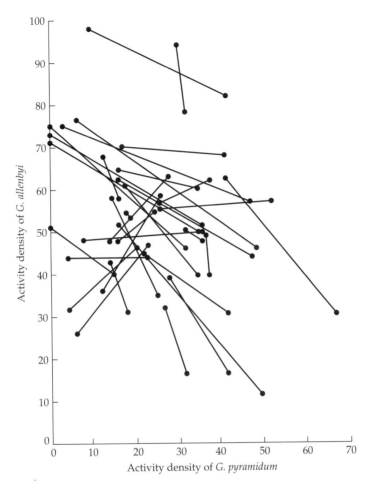

Figure 5.12 Results of gerbil competition experiments plotted in state space. Each line segment connects the points for the high-density and low-density plots in an experimental enclosure. Most of the segments have a negative slope, indicating a reduction in the activity density of *Gerbillus allenbyi* (*y* axis) in the presence of its competitor *G. pyramidum* (*x* axis).

Each line segment in this graph represents the activity density in the two halves of an enclosure. The slope of this line segment is a measure of the isocline of *G. allenbyi* in that area of the state space (Figure 5.12). Although there is considerable scatter in the data, most of these segments have a negative slope, indicating that high densities of *G. pyramidum* depressed the activity of *G. allenbyi*.

Figure 5.13 shows the isocline based on the "best fit" of all these line segments. In contrast to the predictions of the Lotka–Volterra model, the isocline for *G. allenbyi* is nonlinear, with steep declines at high and low densities of *G. pyramidum*, but a shallow slope at intermediate competitor densities.

Why isn't the isocline of *G. allenbyi* a straight line? The answer is that activity density depends not only on the abundance of competitors, but also on the availability and use of different habitats. In the Negev Desert, there are two habitat types that the gerbils use. "Semistabilized dunes" contain little perennial vegetation, many open patches of sand, and unstabilized sand dunes. "Stabilized sand" habitats are dominated by dense shrub cover, with large areas of stable soil crust and few open patches. Both habitat types were present in approximately equal abundance within each enclosure.

Under uncrowded conditions, both gerbil species preferred the semistabilized dunes. As intraspecific density increased, both species began to use the stabilized sand in greater frequency. *G. pyramidum* density induced a habitat shift in *G. allenbyi*, and this was responsible for the nonlinear isocline. Superimposed on the state space in Figure 5.13 are four lines ("isolegs") that are cutpoints for changes in habitat use of the two species. At low densities (regions I and II), both species preferred the semistabilized dunes, and increased densities of *G. pyramidum* led to a sharp decrease in the activity density of *G. allenbyi*. As the density of *G. pyramidum* increased, *G. allenbyi* did not decrease its activity, but instead shifted into the less preferred stabilized sand habitat. Consequently, the isocline is relatively flat in this region, reflecting habitat shift, rather than a reduction in activity density. But as its density increased, *G. pyramidum* was also forced to use the stabilized sand habitat. At high densities of *G. pyramidum*, *G. allenbyi* could no longer "escape" competition by moving to an unoccupied habitat, so activity density again dropped off steeply. Additional field experiments measured the isocline of *G. pyramidum* (Abramsky et al. 1994), and a mathematical analysis predicts stable coexistence of both competitors.

The Lotka–Volterra model generates simple predictions and provides a framework for field tests of competition. Nevertheless, it is very difficult to manipulate species densities in realistic field experiments, and it is still an open question as to whether resources are limiting. These studies show that even when resources are limiting, the model's simple predictions may fail because factors such as variable environments and habitat selection can also affect the outcome of interspecific competition.

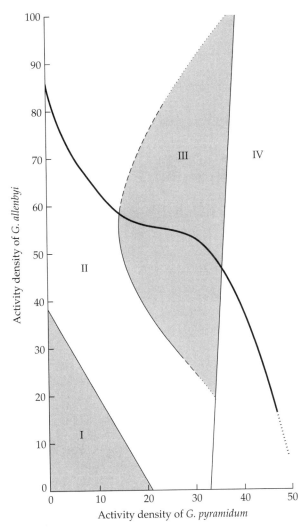

Figure 5.13 The isocline for *Gerbillus allenbyi*, estimated from the data in Figure 5.12. Note that the isocline (thick line) is not linear, but has a region of shallow slope at intermediate densities of *G. pyramidum*. This nonlinear isocline reflects the effects of competition and habitat selection. The thin lines divide the state space into regions based on habitat use. Region I: Both species use the preferred habitat, semistabilized dune. Region II: *G. allenbyi* is forced to use the less preferred habitat, stabilized sand. Region III: Increased use of the stabilized sand by *G. allenbyi*. Because *G. allenbyi* shifts to its less preferred habitat, its activity density can remain high, leading to a shallow slope for the isocline in this region of the state space. Region IV: *G. pyramidum* is forced into the stabilized dune habitat by intraspecific competition. Because *G. allenbyi* no longer has an escape to the unoccupied habitat, its activity density drops off sharply with increases in the activity density of *G. pyramidum*.

Problems

5.1. You are studying competition between red and black desert scorpions. For the red scorpion, $K_1 = 100$ and $\alpha = 2$. For the black scorpion, $K_2 = 150$ and $\beta = 3$.

Suppose the initial population sizes are 25 red scorpions and 50 black scorpions. Graph the state space and isoclines for each species, and plot these initial population sizes. Predict the short-term dynamics of each population and the final outcome of interspecific competition.

5.2. Suppose that, for two competing species, $\alpha = 1.5$, $\beta = 0.5$, and $K_2 = 100$. What is the minimum carrying capacity for species 1 that is necessary for coexistence? How large is the carrying capacity needed for species 1 to win in competition?

*5.3. Diagram the state space for two competing species in which there is a stable equilibrium. Show how intraguild predation could shift this to exclusion by the predatory species.

* Advanced problem

CHAPTER 6

Predation

Model Presentation and Predictions

Competitive interactions in nature are often indirect and subtle, and may be mediated through populations of resources. In contrast, predation is a direct and conspicuous ecological interaction. The image of a wolf pack bringing down a moose, or a spider eating a fly evokes Tennyson's description of "nature red in tooth and claw." Seed predators, such as finches and harvester ants, are less dramatic in their feeding, but may be equally effective at controlling plant populations. Other animals do not consume their prey entirely. Parasites require that their hosts survive long enough for the parasite to reproduce, and many herbivores graze on plants without killing them. In all of these interactions, we can recognize a population of "predators" that benefits from feeding, and a population of "victims" that suffers. In this chapter, we will develop some simple models to give us insight into the dynamics of predation. As in our analysis of competition, the predation equations were first derived independently by Alfred J. Lotka and Vito Volterra. Volterra's interest in the subject stemmed from his daughter's fiancé, a fisheries biologist who was trying to understand fluctuations in the catch of predaceous fish (Kingsland 1985).

MODELING PREY POPULATION GROWTH

We will use the symbol P to denote the predator population and the symbol V to denote the victim or prey population. The growth of the victim population will be some function, f, of the numbers of both victims and predators:

$$\frac{dV}{dt} = f(V, P)$$

Expression 6.1

Suppose that the predators are the only force limiting the growth of the victim population. In other words, if the predators are absent, the victim population increases exponentially:

$$\frac{dV}{dt} = rV$$

Expression 6.2

with r representing the intrinsic rate of increase (see Chapter 1). This potential for increase of the victim population is offset by losses that occur when predators are present:

$$\frac{dV}{dt} = rV - \alpha VP$$

Equation 6.1

The term after the minus sign says that losses to predation are proportional to the *product* of predator and victim numbers. This is equivalent to a chemical reaction in which the reaction rates are proportional to the concentrations of molecules. If predators and victims move randomly through the environment, then their encounter rate is proportional to the product of their abundances. Note that we have now started recycling symbols: α is *not* the competition coefficient from Chapter 5! Instead, here α measures **capture efficiency**, the effect of a predator on the per capita growth rate $\left(\dfrac{1}{V} \dfrac{dV}{dt} \right)$ of the victim population.* The units of α are [victims/(victim • time • predator)]. The larger α is, the more the per capita growth rate of the victim population is depressed by the addition of a single predator. A filter-feeding baleen whale would have a large α, because a single whale can consume millions of plankton. In contrast, a web-building spider might have a fairly low α if the addition of a single web does not greatly depress prey populations. The product αV is the **functional response** of the predator—the rate of victim capture by a predator as a function of victim abundance (Solomon 1949). Later in this chapter, we will derive some more complicated expressions for the functional response, but for now we will represent it is as a simple product of victim abundance (V) and capture efficiency (α). Before we explore the solutions to the equation for victim growth, we will develop an analogous equation to describe the growth of the predator population.

MODELING PREDATOR POPULATION GROWTH

The growth of the predator population is affected by the numbers of both predators and victims:

$$\frac{dP}{dt} = g(P,V) \qquad\qquad \text{Expression 6.3}$$

We use the symbol g for this function to distinguish it from the function f that is used for the victim population in Expression 6.1.

The predator we are modeling is an extreme specialist. It will feed only on the victim population and has no alternative source of prey. Consequently, if the victim population is absent, the predator population declines exponentially:

$$\frac{dP}{dt} = -qP \qquad\qquad \text{Expression 6.4}$$

where q is the per capita **death rate**. (This is equivalent to the death rate d from the exponential growth model described in Chapter 1; we have changed symbols here to avoid confusion.)

---

*This same capture efficiency appeared as the interaction coefficient δ in Equation 5.8, where it represented losses to predation in a pair of competitors engaging in intraguild predation.

Positive growth occurs only when the victim population is present:

$$\frac{dP}{dt} = \beta VP - qP$$

Equation 6.2

Here βVP indicates random encounters of predators and victims. β is a measure of **conversion efficiency***—the ability of predators to convert each new victim into additional per capita growth rate for the predator population $\left(\frac{1}{P}\frac{dP}{dt}\right)$. Its units are [predators/(predator • time • victim)]. We expect β to be high when a single prey item is particularly valuable, such as a moose that is captured by wolves. In contrast, β will be low when a single prey item does not contribute much to growth of the predator population; think of a single seed consumed by a granivorous bird. βV reflects the **numerical response** of the predator population—the per capita growth rate of the predator population as a function of victim abundance.

EQUILIBRIUM SOLUTIONS

To find the equilibrium for the victim and predator populations, we set each equation equal to zero and solve for population size. Beginning with Equation 6.1:

$$0 = rV - \alpha VP$$

Expression 6.5

$$rV = \alpha VP$$

Expression 6.6

$$r = \alpha P$$

Expression 6.7

$$\hat{P} = \frac{r}{\alpha}$$

Equation 6.3

Although we tried to solve for the victim equilibrium, the solution is in terms of P, the predator population! The important result is that a specific number of predators ($\hat{P}$) will maintain the victim population at zero growth. This predator level is determined by the ratio of the growth rate of the victims (r) to the capture efficiency of the predators (α). The faster the growth rate of the

---

*Again, this conversion efficiency appeared as the interaction coefficient γ in Equation 5.7 of Chapter 5, where it represented gains from predation in a pair of competitors engaging in intraguild predation.

victim population, the more predators are needed to keep the victim population in check. Conversely, the higher the capture efficiency, the fewer predators needed for control.

Solving the equilibrium for the predators (Equation 6.2) yields an expression in terms of the victim population size:

$$0 = \beta VP - qP \qquad \text{Expression 6.8}$$

$$\beta VP = qP \qquad \text{Expression 6.9}$$

$$\beta V = q \qquad \text{Expression 6.10}$$

$$\hat{V} = \frac{q}{\beta} \qquad \text{Equation 6.4}$$

Thus, the predator population is controlled by a fixed number of victims ($\hat{V}$). The greater the death rate of the predators (q), the more victims needed to keep the predator population from declining. Conversely, the greater the conversion efficiency of predators (β), the fewer victims needed to maintain the predators at equilibrium. Because Equations 6.3 and 6.4 give the conditions for zero growth, they represent the victim and predator isoclines, respectively.

GRAPHICAL SOLUTIONS TO THE LOTKA–VOLTERRA PREDATION MODEL

As in our analysis of the competition model (Chapter 5), we can plot the isoclines for each species in state space to evaluate the joint equilibrium. Plotting the victim population on the x axis yields a horizontal victim isocline, representing the number of predators needed to hold the victim population in check. If the predator population is less than this number, the victim population can increase in size, represented by horizontal arrows pointing to the right. Conversely, if the predator population is above the victim isocline, the victim population declines, represented by horizontal arrows pointing to the left (Figure 6.1).

Similar reasoning applies to the analysis of the predator isocline. This isocline is a vertical line, representing a critical size of the victim population. To the left of the isocline, there are not enough victims to support the predator population. In this region of the state-space graph, the predator population declines, represented by downward-pointing vertical arrows. To the right of the isocline, there is an excess supply of victims, and the predator population increases, represented by upward-pointing vertical arrows (Figure 6.2).

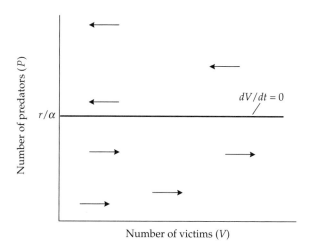

Figure 6.1 The victim isocline in state space. The Lotka–Volterra predation model predicts a critical number of predators (r/α) that controls the victim population. If there are fewer predators than this, the victim population increases (right-pointing arrows). If there are more predators, the victim population decreases (left-pointing arrows). The victim population has zero growth when $P = r/\alpha$.

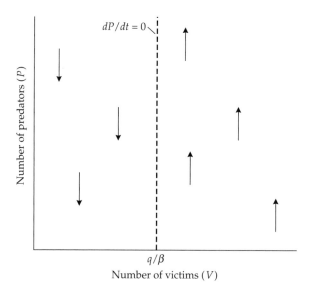

Figure 6.2 The predator isocline in state space. The Lotka–Volterra predation model predicts a critical number of victims (q/β) that controls the predator population. If there are fewer victims than this, the predator population decreases (downward-pointing arrows). If there are more victims, the predator population increases (upward-pointing arrows). The predator population has zero growth when $V = q/\beta$.

In our analysis of competition models, there were four ways that the pair of isoclines could be placed in the state-space graph. For the predation model, there is only one possible pattern: the isoclines cross at 90° angles (Figure 6.3). However, we will see that the dynamics are more complex than in the competition model.

The predator and victim isoclines divide the state space into four regions. Beginning in the upper right-hand corner, we are in a region where both predator and victim are abundant. Because we are to the right of the predator isocline, prey are abundant enough for the predator to increase. However, we are above the horizontal victim isocline. Consequently, there are too many predators, and the victim population declines. The vector of net movement points towards the upper left-hand quadrant. As the victim abundance continues to decline, we cross the vertical isocline into the upper left-hand region of state space.

Now the victim population has declined to the point where the predator population can no longer increase. Both predator and victim populations decrease, and the vector moves into the lower left-hand quadrant. In this region, the predator population continues to decline, but the victim population starts to increase again. The net movement is down and to the right, taking the trajectory into the fourth quadrant. Here, the victim population con-

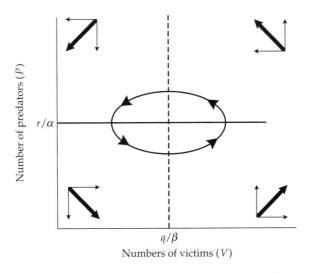

Figure 6.3 The dynamics of predator and victim populations in the Lotka–Volterra model. The vectors indicate the trajectories of the populations in the different regions of the state space. The populations trace a counterclockwise path that approximates an ellipse.

tinues to grow, but it has now become large enough for the predators to also increase. The system again moves back to the starting point, the upper right-hand quadrant.

Thus, the predator and victim populations trace an approximate ellipse in state space. Unless the predator and victim populations are precisely at the intersection of the isoclines, their trajectories will continue to move in this counterclockwise ellipse.

How does this ellipse translate into growth curves for the predator and victim populations? Both populations cycle periodically, increasing and decreasing smoothly from minimum to maximum. The ellipse indicates that the peak of the predator population occurs when the victim population is at its midpoint, and vice versa. In other words, the peaks of the predator and victim populations are displaced by one-quarter of a cycle (Figure 6.4).

What would happen if the predator and victim populations had a different starting point in the state space? This would correspond to different initial abundances of predator and victim, and a new ellipse would be traced. Both populations would again exhibit cycles, although with a different amplitude. The closer the ellipse is to the isocline intersection, the smaller the amplitude of the predator and victim cycles. Thus, the Lotka–Volterra cycles are neutrally stable—the amplitudes are determined solely by the initial conditions.

There are only two exceptions to population cycling: (1) if the victim and predator populations are precisely at the isocline intersection, they will not

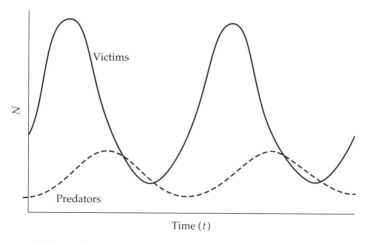

Figure 6.4 Cycles of predators and victims in the Lotka–Volterra model. Each population cycles with an amplitude that is determined by the starting population sizes and a period of approximately $2\pi/\sqrt{rq}$. The predator and victim populations are displaced by one-quarter of a cycle, so that the predator population peaks when the victim population has declined to half its maximum, and vice versa.

change, although if they are displaced any distance from this point, they will begin cycling; or (2) if the starting point of the ellipse is too extreme, it will hit one of the axes of the state-space graph. In this case, the amplitude of the cycle is so large that either predator or the victim population will crash. Although the amplitude of the cycle is determined by the initial population sizes, the period of the cycle (C) is approximately

$$C \approx \frac{2\pi}{\sqrt{rq}}$$ Equation 6.5

Thus, the greater the prey growth rate (r) and/or the predator death rate (q), the faster the populations cycle between high and low values. The essential feature of the Lotka–Volterra predation model is that the predator and victim populations cycle because they reciprocally control one another's growth.

Model Assumptions

The Lotka–Volterra predation model carries with it the standard assumptions of no immigration, no age or genetic structure, and no time lags. In addition, the model makes the following assumptions about predators, victims, and the environment:

✔ **Growth of the victim population is limited only by predation.** Equation 6.1 shows that the victim population grows exponentially in the absence of the predator.

✔ **The predator is a specialist that can persist only if the victim population is present.** Equation 6.2 shows that the predator population will starve in the absence of the victim.

✔ **Individual predators can consume an infinite number of victims.** Because the horizontal victim isocline ($dV/dt = 0$) implies a constant number of predators, each predator must be able to increase its consumption as the victim population increases in size. An infinite capacity for consuming prey also implies that there is no interference or cooperation among predators.

✔ **Predator and victim encounter one another randomly in an homogenous environment.** The interaction terms (αVP and βVP) imply that predators and victims move randomly through the environment, and that victims do not have spatial or temporal refuges for avoiding predators.

Model Variations

The unique prediction of the Lotka–Volterra predation model is cycles of predator and victim populations. However, these cycles are very sensitive to the restrictive assumptions and linear isoclines of the model. In the following sections, we will incorporate more realistic assumptions about predators and victims that bend the isoclines and produce other dynamics. We will not solve the equations for these more complex models, although we will analyze their behavior with state-space graphs.

INCORPORATING A VICTIM CARRYING CAPACITY

The victim isocline tells us how many predators are needed to hold the victim population in check. Notice that as we move to the right in the state-space graph (Figure 6.1), the same number of predators will control the victim population. This is not realistic. We expect that as the victim population becomes more crowded, it will start to be limited by other resources that have nothing to do with predators. We can modify the victim isocline to incorporate a victim carrying capacity by including another term with a new constant c:

$$\frac{dV}{dt} = rV - \alpha VP - cV^2 \qquad \text{Equation 6.6}$$

Now the growth of the victim population is decreased by the presence of predators (αVP) and by its own abundance (cV^2). When graphed in the state space, this new isocline is a straight line with a negative slope, in contrast to the horizontal victim isocline of the simple Lotka–Volterra model. The new isocline crosses the x axis at r/c, which is the maximum population size achieved by the victims when no predators are present. In the absence of predators, Equation 6.6 is equivalent to a model of logistic population growth with a carrying capacity $K = r/c$ (Equation 2.1).

How does the interaction of predator and victim change when the victim population is limited by its own carrying capacity? Figure 6.5 shows that the trajectory for the predator and victim populations spirals inwards to the equilibrium intersection. This is a stable equilibrium point, and the equilibrium abundance for the victim population is lower when the predators are present than when they are absent. The presence of a victim carrying capacity stabilizes the predator–prey interaction. This makes intuitive sense—if the victims are limited by factors other than their predators, then there would be less of a tendency for the two populations to cycle.

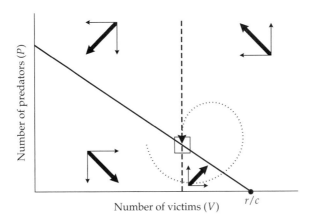

Figure 6.5 The effect of a victim carrying capacity on the victim isocline. The victim isocline slopes downward with a carrying capacity incorporated. The intersection with the vertical predator isocline forms a stable equilibrium point.

MODIFYING THE FUNCTIONAL RESPONSE

One of the most unrealistic assumptions of the Lotka–Volterra predation model is that individual predators can always increase their prey consumption as the victim population increases. This type of foraging is illustrated in a graph of the functional response (Figure 6.6), which plots the rate of prey

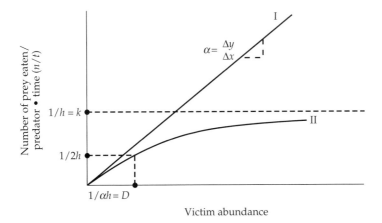

Figure 6.6 The functional response of predators is the feeding rate per predator as a function of prey abundance. The shape of these curves depends on the capture efficiency (α), the maximum predator feeding rate (k), and the victim abundance for which the predator feeding rate is half of the maximum (D).

captured per individual predator (n/t) as a function of prey abundance (V). The Lotka–Volterra model assumes a **Type I functional response**, in which the predator consumes more as prey abundance increases (Holling 1959). The slope of this curve is α, the capture efficiency.

A Type I functional response is unrealistic for two reasons. First, predators will eventually become satiated (stuffed) and stop feeding. Second, even in the absence of satiation, predators are limited by the **handling time** (h) needed to catch and consume each prey item. Consequently, there is a limit to the rate at which individual predators can process prey.

We can construct a more realistic **Type II functional response** by modeling the components that contribute to **feeding rate** (n/t), the rate at which individual predators capture prey (Royama 1971). The total amount of time that a predator spends feeding (t) is the time spent searching for the prey (t_s), plus the time spent "handling" or consuming the prey (t_h):

$$t = t_s + t_h \qquad \text{Expression 6.11}$$

If we let n equal the number of prey items captured in time t and h equal the handling time per prey item, the total handling time is:

$$t_h = hn \qquad \text{Expression 6.12}$$

Similarly, we can derive an expression for the search time. The total number of prey captured by a predator (n) is simply the product of the victim abundance (V), the capture efficiency (α), and the total search time (t_s):

$$n = V\alpha t_s \qquad \text{Expression 6.13}$$

We can rearrange this to give us an expression for the search time:

$$t_s = \frac{n}{\alpha V} \qquad \text{Expression 6.14}$$

Substituting Expressions 6.12 and 6.14 into 6.11, we have:

$$t = \frac{n}{\alpha V} + hn \qquad \text{Expression 6.15}$$

Multiplying the second term by ($\alpha V / \alpha V$) gives:

$$t = \frac{n}{\alpha V} + \frac{\alpha V h n}{\alpha V} \qquad \text{Expression 6.16}$$

$$t = n\left(\frac{1 + \alpha V h}{\alpha V}\right) \qquad \text{Expression 6.17}$$

Finally, this can be rearranged to give us an expression for the feeding rate (n/t):

$$n/t = \frac{\alpha V}{1 + \alpha V h} \qquad\qquad \text{Equation 6.7}$$

Equation 6.7 describes the feeding rate per predator as a function of the capture efficiency, the victim abundance, and the handling time. Note that if the victim abundance is very low, the term $\alpha V h$ in the denominator is small, so the feeding rate is close to αV, as in the simple Lotka–Volterra model. But as the victim abundance increases, the feeding rate approaches a saturation value of $1/h$. This value represents the maximum feeding rate that the predator can achieve because of the constraints of handling time. Equation 6.7 is sometimes referred to as the "disc equation" because it fits data from an experiment in which human subjects were blindfolded and required to find and pick up small discs of sandpaper scattered on a flat surface.

We can simplify Equation 6.7 somewhat by letting $k = 1/h$, the **maximum feeding rate**. We can also define the constant D as $1/\alpha h$. This value turns out to be the **half-saturation constant**, which is the abundance of prey at which the feeding rate is half-maximal. If we first multiply the numerator and denominator of Equation 6.7 by $1/\alpha h$, we have:

$$n/t = \frac{\dfrac{\alpha V}{\alpha h}}{\dfrac{1}{\alpha h} + \dfrac{\alpha V h}{\alpha h}} \qquad\qquad \text{Expression 6.18}$$

Substituting in the two new constants k and D yields:

$$n/t = \frac{kV}{D + V} \qquad\qquad \text{Equation 6.8}$$

This Type II functional response increases to a maximum and constant rate of prey consumption per predator (k). The half-saturation constant (D) controls the rate of increase to this maximum. This equation is identical to the Michaelis–Menten equation of enzyme kinetics (Real 1977).

Finally, a **Type III functional response** can be described by:

$$n/t = \frac{kV^2}{D^2 + V^2} \qquad\qquad \text{Equation 6.9}$$

For a Type III functional response, the feeding rate also reaches an asymptote at k, but the curve has a sigmoid shape, similar to the logistic curve (see Chapter 2). Consequently, the feeding rate is accelerated at low prey density, but decreases at high prey density as the asymptote is reached (Figure 6.7).

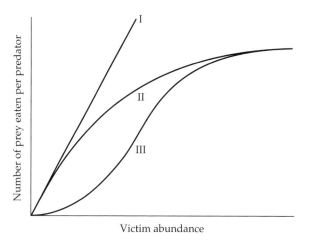

Figure 6.7 Type I, Type II, and Type III functional responses.

This functional response can occur if predators switch to prey items that become more common, if they develop a search image that increases capture efficiency as victim abundance increases, or if there are fixed and variable costs to foraging (Holling 1959, Mitchell and Brown 1990).

The functional response has important consequences for the ability of predators to control victim populations. Figure 6.8 shows the proportion of the prey population that is consumed by an individual predator as victim abundance increases. For the Type I response of the simple Lotka–Volterra model, this proportion remains a constant, because each predator increases its

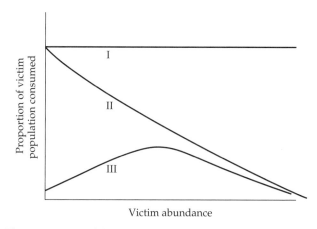

Figure 6.8 The proportion of the victim population consumed by an individual predator as a function of victim abundance.

individual feeding as victim abundance increases. For the Type II response, the proportion decreases steadily because each predator can only process prey at a maximum rate k. The Type III response shows an initial increase because of the accelerated feeding rate, but this quickly decreases and converges on the Type II curve. These curves show that, at high victim abundance, predators with a Type II or Type III response may not be able to effectively control victim populations. Control is possible with the Type III response, but only at relatively low victim abundance. In contrast, the Type I functional response ensures effective control over all levels of victim abundance.

Incorporating a Type II or Type III functional response into the equation for the victim growth rate gives:

$$\frac{dV}{dt} = rV - \left(\frac{kV}{V+D}\right)P \qquad \text{Equation 6.10}$$

$$\frac{dV}{dt} = rV - \left(\frac{kV^2}{V^2+D^2}\right)P \qquad \text{Equation 6.11}$$

Figure 6.9 shows that the isoclines for these growth equations increase in the state space, with an upward swing at low victim abundance for the Type III

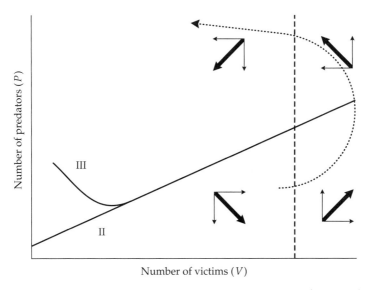

Figure 6.9 Victim isoclines incorporating a Type II or a Type III functional response. The intersection of an increasing victim isocline with a vertical predator isocline generates an unstable equilibrium point.

functional response. Because each predator is limited by a maximum consumption rate, more predators are required to hold large victim populations at zero growth. When these increasing victim isoclines intersect a vertical predator isocline, the equilibrium is unstable, and the predator and victim will not coexist.

THE PARADOX OF ENRICHMENT

The victim isocline may also increase because of an Allee effect (see Chapter 2) for the victim population. If larger victim populations are more effective at reproducing, obtaining food, or defending themselves from predators, more predators would be needed to control the prey population. Because of a victim carrying capacity, predator functional response, Allee effects, and a variety of other reasons, the victim isocline may have a hump in the middle (Rosenzweig and MacArthur 1963), turning downward at both low and high prey densities.

How does this more realistic victim isocline affect predator–prey dynamics? The answer depends on precisely where the vertical predator isocline intersects the victim isocline. If the intersection is at the peak of the victim isocline, the predator and victim populations will cycle as in the simple Lotka–Volterra model (Figure 6.10a). However, if the predator isocline crosses to the right of the hump, the predator and victim populations converge on a stable equilibrium point, without population cycles (Figure 6.10b). In this case, the predator is relatively inefficient. Thus, from Equation 6.4, the predator population has a relatively high death rate (q) and/or a low conversion efficiency (β). In contrast, if the predator is relatively efficient (low q and/or high β), the isoclines intersect to the left of the hump. In this case, the equilibrium is unstable. The predator population will overexploit the victim population, drive it to extinction, and then starve (Figure 6.10c).

This instability due to a relatively efficient predator has been termed the **paradox of enrichment** (Rosenzweig 1971). The paradox may explain why some artificially enriched agricultural systems are vulnerable to pest outbreaks. Suppose the "victim" population is a crop plant that coexists in a stable equilibrium with a "predator" population of an herbivorous insect. If the productivity of the crop plant is increased with fertilizers, the victim isocline

Figure 6.10 (a) Predator–prey cycles with a humped prey isocline. As in the Lotka–Volterra model, the predator and victim populations cycle as long as the predator and victim isoclines are perpendicular where they intersect. (b) If the predator is relatively inefficient, the predator isocline intersects to the right of the peak of the victim isocline. In this case, predator and victim coexist in a stable equilibrium. (c) If the predator is relatively efficient, the predator isocline intersects to the left of the peak of the victim isocline. In this case, the predator overexploits the prey population, drives it to extinction, and starves.

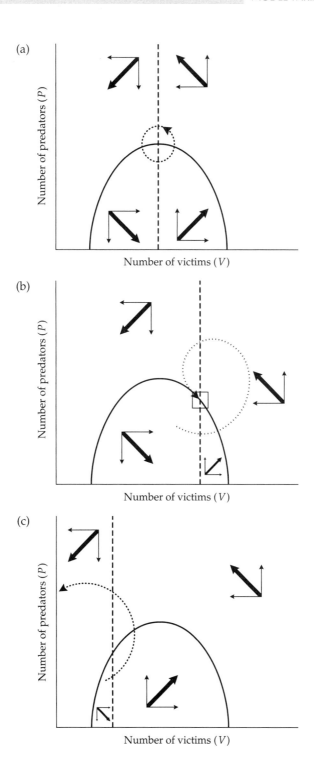

(a)

Number of predators (P)

Number of victims (V)

(b)

Number of predators (P)

Number of victims (V)

(c)

Number of predators (P)

Number of victims (V)

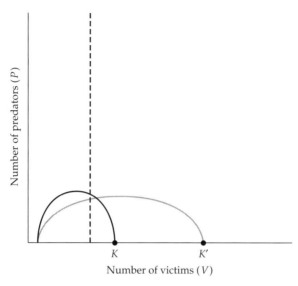

Figure 6.11 The paradox of enrichment. If the victim population has its carrying capacity enhanced from K to K', the system moves from a stable equilibrium to overexploitation by the predator.

may shift to the right to a new, higher carrying capacity (Figure 6.11). If the predator isocline remains stationary, the dynamics may shift from a stable equilibrium to an unstable outbreak of the "pest." This paradox depends on the unrealistic assumption of a strictly vertical predator isocline. More realistic predator isoclines, described later in this chapter, may enhance stability of predator and prey over a wider range of victim abundances (Berryman 1992).

INCORPORATING OTHER FACTORS IN THE VICTIM ISOCLINE

The victim isocline may also turn upward at low victim abundance, generating different population dynamics. There are at least three reasons for an upturn of the victim isocline. First, the isocline will turn up if there is a fixed number of victim refuges that are secure from predators. For example, fish that live in rock crevices and songbirds that establish territories in areas protected by cover have spatial refuges from predation. In this case, no matter how large the predator population gets, the victim population can always persist at low abundance in the refuges. Second, the victim isocline may turn upwards if there is a constant number of victim immigrants that arrive each generation. With immigration, the victim population always has the potential to increase at low numbers. Finally, the isocline may turn upward at low victim abundance because of a Type III functional response, as explained earlier.

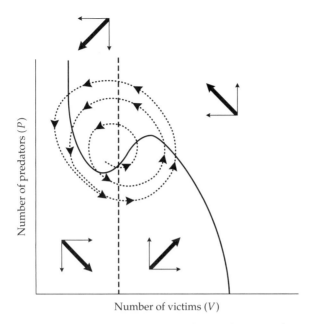

Figure 6.12 Cycling of predator and victim populations because of victim refuges. If there are spatial refuges from predation, the victim isocline becomes vertical at low victim abundance. In this case, the efficient predator cannot overexploit its prey, and begins to starve once all the victims outside of the refuges have been consumed. After the predator population declines below a certain point, the victim population begins to increase again, repeating the cycle.

This upward turn of the victim isocline has the potential to stabilize predator–prey interactions. For example, suppose that the predator is relatively efficient, but there is a victim carrying capacity and there are refuges from predation for the victim population (Figure 6.12). In this case, the predators quickly consume all the available victims, as in the destabilized case (Figure 6.10c). But once all the victims outside the shelters are consumed, the predator population begins to starve, and its abundance declines. When the predator population declines below a certain point, the victim population in the refuges starts to increase, and the cycle repeats itself. In contrast to the simple Lotka–Volterra model, these cycles are stable, because no matter what the starting density, the predator population will eventually consume all the victims not in refuges, and the cycle will repeat.

MODIFYING THE PREDATOR ISOCLINE

We can also modify the vertical predator isocline to make it more realistic. These modifications involve changes in the numerical response of Equation

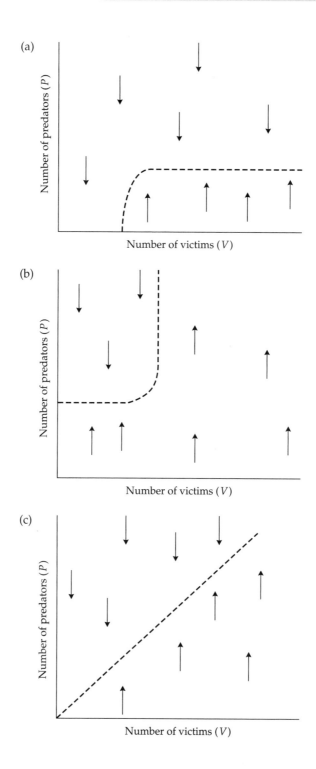

◀ **Figure 6.13** (a) Effects of carrying capacity on the predator isocline. If the predator population is limited by factors other than victim abundance, the predator isocline bends to the right. No matter how large the victim population, the predator population becomes limited when it reaches its own carrying capacity. (b) Effects of the availability of alternative prey on the predator isocline. If the predator is not a specialist on the victim, the predator population may be able to increase even when the victim abundance declines to zero. (c) Effects of victim abundance on the predator isocline. If the size of the victim population acts as a carrying capacity for the predators, the predator isocline increases with increasing victim abundance.

6.2, which will be described qualitatively. For example, the Lotka–Volterra predation model assumes that the predator population can always increase in size if there is an excess of prey available. It is more realistic to suppose that the predator population has its own carrying capacity, so that its growth is limited by other factors. A carrying capacity for the predator bends the predator isocline to the right (Figure 6.13a).

Another unrealistic assumption of the Lotka–Volterra model is that the predator is a specialist on the victim. Suppose instead that the predator has alternative prey sources. Then, when the victim population becomes less abundant, the predator population can continue to increase by feeding on other prey items. This will tip the predator isocline towards the horizontal at low prey abundance (Figure 6.13b). Thus, with alternative prey and a predator carrying capacity, the predator isocline can shift from vertical to horizontal. As we noted earlier, the availability of other prey may shift the victim isocline as well.

As an intermediate case, suppose that the size of the victim population determines the size of the predator population. In other words, the victim population functions as a "carrying capacity" for the predators. In this case, the predator isocline will be a line with a positive slope, intermediate between the vertical isocline of the Lotka–Volterra model and the horizontal isocline of a predator with an independent carrying capacity and alternative prey (Figure 6.13c).

How will these alterations of the predator isocline affect the stability of the model? As a general rule, anything that rotates either the predator or the victim isocline in a *clockwise* direction will tend to stabilize the interaction, whereas anything that rotates the isoclines *counterclockwise* will be destabilizing. These rotations can be compared to the neutral stability of a horizontal victim isocline and a vertical predator isocline in the Lotka–Volterra model (Figure 6.14). For example, giving the victim population a carrying capacity rotates the victim isocline clockwise, leading to a stable equilibrium on the right side of the hump (Figure 6.10b). But adding predator satiation rotates the victim isocline counterclockwise at low abundances, leading to an unstable equilibrium on the left side of the hump (Figure 6.10c). Rotating the

Decreasing stability   Increasing stability

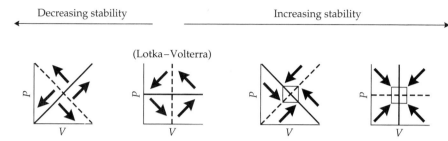

Figure 6.14 Effects of rotating the predator and victim isoclines on the stability of the equilibrium. Relative to the neutral equilibrium of the Lotka–Volterra model, clockwise rotations of the isoclines lead to more stable equilibria; counterclockwise rotations lead to less stable equilibria.

predator isocline also increases the stability of the interaction. Whereas a vertical predator isocline generates population cycles in a neutral equilibrium, an increasing predator isocline generates damped cycles, and a horizontal predator isocline generates a stable equilibrium point (Figure 6.15).

These geometrical rules make intuitive biological sense. The more independent the predator and prey are of one another, the more stable the inter-

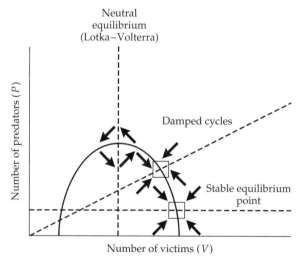

Figure 6.15 Effects of clockwise rotation of the predator isocline. As the predator isocline is rotated, the dynamics change from cycles with a neutral equilibrium, to damped cycles, to a stable equilibrium point. Biologically, the three predator isoclines correspond to a predator that is a complete specialist on the victim, to one whose carrying capacity is proportional to victim abundance, to one whose carrying capacity is independent of victim abundance.

action. For example, suppose the victim isocline is vertical and the predator isocline is horizontal (Figure 6.14). In this case, the carrying capacities of the predator and victim are completely independent of one another, and both species coexist in a very stable equilibrium. Cycles are difficult to generate with simple predator–victim models, and require a special dependence of predator and victim populations upon each other, as in the original Lotka–Volterra model.

Empirical Examples

POPULATION CYCLES OF HARE AND LYNX

The basic prediction of the Lotka–Volterra model is the regular cycling of predator and prey populations. The most famous example of this cycling is the case of the Canada lynx (*Lynx canadensis*) and its principal prey, the snowshoe hare (*Lepus americanus*). The ecologist Charles Elton analyzed fur-trapping records from the Hudson's Bay Company in Canada and found a long-term record of population cycles (Elton and Nicholson 1942). The major source of hare mortality is predation (Smith et al. 1988), and the hare population cycles with a peak abundance approximately every 10 years (Figure 6.16). The lynx population is highly synchronized with the hare and usually peaks one or two years later. These are not the only prey and predator species that cycle in the boreal north. Populations of muskrat, ruffed grouse, and ptarmigan exhibit 9 to 10 year cycles, whereas smaller herbivores such as voles and lemmings cycle with peaks every 4 years. Predators such as foxes, mink, owls, and martens also cycle synchronously with their prey.

What is the explanation for the striking hare–lynx cycle? An early suggestion that the hare cycles were correlated with sunspot activity was dismissed

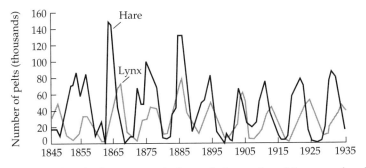

Figure 6.16 One-hundred-year record of population cycles of the snowshoe hare (*Lepus americanus*) and the Canada lynx (*Lynx canadensis*), based on pelt records of the Hudson's Bay Company in Canada.

because sunspot activity peaks every 11 years, whereas the hare cycle is approximately 10 years in length (Moran 1949). For many years, the hare–lynx cycle was the classic textbook example of predator and prey populations that cycled according to the Lotka–Volterra model. More recently, ratio-dependent predator–prey models have been applied to the hare–lynx data (Akçakaya 1992). These models are based on the assumption that the functional response of the predator depends not simply on victim abundance (V), but on the ratio of prey to predator abundance (V/P) (Arditi and Ginzburg 1989).

Unfortunately, two additional pieces of data complicate the story. First, the hare–lynx cycles seem to be broadly synchronized within a year or two over wide areas of North America (Smith 1983). If the predator–prey models were correct, we would expect cycles of different amplitude and period to arise in different local populations. Second, there are places on the coast of British Columbia and on Anticosti Island, Quebec, where there are no lynx, but the hare population cycles nonetheless!

These results suggest that the hare and lynx do not reciprocally influence each other. Instead, the lynx population is probably "tracking" the hare cycle. The hare cycle seems to be caused, in part, by interactions with its food supply. Heavily grazed grasses produce shoots with high levels of toxins that make them less palatable to hares (Keith 1983). This chemical protection persists for two or three years after grazing, further contributing to the hare decline. A single-species logistic model with a time lag (see Chapter 2) would qualitatively describe this sort of cycle. However, as most hares die of predation, not starvation, food quality probably contributes to their susceptibility to predation.

Finally, recent evidence again suggests that sunspots may indeed contribute to the cycles. Sunspot activity is associated with hare browse marks in tree rings and with periods of low snow accumulation (Sinclair et al. 1993). Sunspot activity may serve as a phase-locking mechanism through indirect influences on climate and plant growth. These broad climatic effects could be responsible for the synchrony of hare–lynx cycles over large areas of Canada and Alaska. However, the degree of synchrony among continents is currently being debated (Ranta et al. 1997; Sinclair and Gosline 1997). Whatever the ultimate explanation, it is clear that the hare-lynx cycle is more complex than suggested by the superficial match of the data to the simple predictions of the Lotka–Volterra model.

POPULATION CYCLES OF RED GROUSE

Interactions between hosts and parasites represent a special kind of "predation" in which the life history of the predator is intimately tied to that of its host. Whereas most predators benefit from rapidly killing and consuming their prey, a parasite must keep its host alive at least long enough to successfully reproduce and infect a new host. To understand the population dynamics of hosts and parasites, we must therefore model the dynamics of the egg or larval stages, as well as those of the host and the adult parasite (Anderson and May 1978).

A nice example illustrating these complexities is the case of the parasitic nematode *Trichostrongylus tenuis*, which infects red grouse (*Lagopus lagopus scoticus*) on the moors of England and Scotland. Adult worms inhabit the large caeca of red grouse, and their eggs pass out of the host with the feces. If the environment is warm and moist, the eggs hatch and develop into a larval stage. The larval nematode moves to the growing tips of heather plants, where it is consumed by a new host, and the life cycle repeats itself. A single bird may be host to over 10,000 worms. As the intensity of the parasite infection increases, winter mortality, egg mortality, and chick losses all increase (Figure 6.17). Thus, *T. tenuis* has the potential to regulate the population growth of red grouse.

Because red grouse are an important game bird in England and Scotland, there are detailed records on its population dynamics and the prevalence of parasite infection (Hudson et al. 1992). Figure 6.18 shows a 14-year record of host and parasite populations at Gunnerside, North Yorkshire. The red grouse population cycles, with a period of approximately 5 years. Parasite burden (number of worms per host) also cycles, with peaks occurring near the low point of the red grouse cycle.

Even a relatively simple model of the grouse–nematode interaction requires a minimum of three differential equations: one for the host (H), one for the adult worms (P), and one for the free-living egg and larval stages (W; Dobson and Hudson 1992). The growth of the host population can be modeled as:

$$\frac{dH}{dt} = (b - d - cH)H - (\alpha + \delta)P \qquad \text{Expression 6.19}$$

The first term $[(b - d - cH)H]$ represents the growth of the red grouse population in the absence of the parasite. The constants b and d represent intrinsic birth and death rates, and cH is a density-dependent term. The first part of this equation is really a model of logistic growth, with a carrying capacity of $[(b - d)/c]$. A finite carrying capacity is realistic for the grouse population because the birds are territorial. The second part of the equation $[(\alpha + \delta)P]$ represents the losses due to parasites. α is the reduction in host population growth due to effects of the parasite on the survivorship of grouse, and δ is the reduction due to parasite effects on the reproduction of grouse. We distinguish between these two mechanisms because α and δ appear separately in other equations in the model.

Next, we write an equation for the growth rate of the free-living stages:

$$\frac{dW}{dt} = \lambda P - \gamma W - \beta WH \qquad \text{Expression 6.20}$$

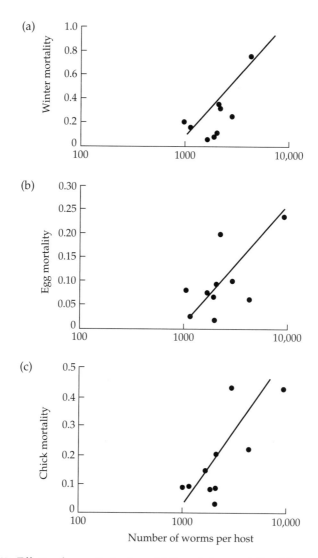

Figure 6.17 Effects of parasite load on (a) the winter mortality, (b) egg mortality, and (c) chick losses of red grouse (*Lagopus lagopus scoticus*). The *x* axis is the average parasite load (worms per host), and the *y* axis is the proportional mortality caused by each factor. Because the nematode *Trichostrongylus tenuis* reduces both the survivorship and reproduction of red grouse, it has the potential to regulate host numbers. (From Hudson et al. 1992.)

Here, λ is the per capita fecundity of the parasite in the host, γ is the death rate of the egg and larval stages in the field, and βWH is the rate at which larvae are transmitted to new hosts. Note the similarity of this latter expression to the

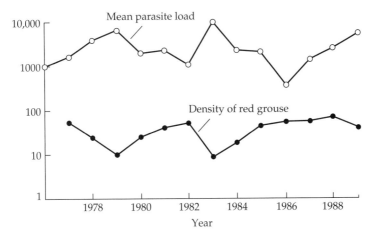

Figure 6.18 Changes in red grouse (*Lagopus lagopus scoticus*) density (breeding hens per square kilometer) and mean parasite load (worms per host) over 14 years at Gunnerside, North Yorkshire. Both the grouse and the nematode populations cycle with a period of approximately 5 years. Note the logarithmic scale on the *y* axis. (From Dobson et al. 1992; data from Hudson et al. 1992.)

"random encounter" term in the Lotka–Volterra model (Equations 6.1 and 6.2). Finally, we can describe the dynamics of the adult worm population as:

$$\frac{dP}{dt} = \beta WH - (\mu + d + \alpha)P - \alpha \frac{P^2}{H}\left(\frac{k+1}{k}\right) \quad \text{Expression 6.21}$$

The first term (βWH) represents the increase in the adult worm population from transmission. This is equivalent to the loss component of the egg-larva population. The second term $[(\mu + d + \alpha)P]$ represents decreases in growth of the worm population due to parasite death (μ), intrinsic host mortality (d), and host mortality from parasitism (α). The final term, $[\alpha(P^2/H)][(k+1)/k]$, represents losses due to the spatial dispersion of the worms among hosts. The constant k describes the distribution of worms among hosts. The smaller k is, the more aggregated the worms are in a few hosts. Aggregation will tend to decrease the growth of the worm population as the few heavily infected hosts die and take their parasites with them! In contrast, if the worms are distributed randomly or evenly among hosts, the growth rate of the parasite population is increased.

With ten different parameters in the model, there are a variety of possible outcomes. If parasite and host fecundity are not high enough, the parasite will go extinct, and the grouse population will rise to its carrying capacity. If the larval life of the parasite is relatively short, the grouse and parasite populations will coexist in a stable equilibrium. But if the larval and egg stages are fairly long-lived, the model generates stable cycles of host and parasite populations. Cycles in this model arise when $\alpha/\delta > k$. In other words, the ratio of parasite effects on survivorship (α) to parasite effects on reproduction (δ) must exceed the degree of parasite aggregation among hosts (k).

Field data were used to independently estimate the parameters of Expressions 6.19–6.21. The resulting model predicted population cycles with a period of approximately five years, which was observed at Gunnerside (Dobson and Hudson 1992). The model also provides insight into other grouse populations in England and Scotland. Not all grouse and nematode populations cycle, and these noncycling populations are in areas of relatively low rainfall (Hudson et al. 1985). Under these circumstances, the survival of eggs and larvae outside of the host is poor, and the model does not predict cycles.

The interaction of red grouse and its nematode parasite is one of the few well-documented cases of a predator and victim that cause each other's populations to cycle. However, the biology of the system is considerably more complex than that described by the simple Lotka–Volterra model. Models of host–parasite interactions have also been used to predict the dynamics of HIV (the AIDS virus) that infects humans.

Problems

6.1. Suppose that spider and fly populations are governed by Lotka–Volterra dynamics, with the following coefficients: $r = 0.1$, $q = 0.5$, $\alpha = \beta = 0.001$. If the initial population sizes are 200 spiders and 600 flies, what are the short-term population dynamics predicted by the model?

6.2. Suppose that hawk and dove populations cycle with a peak every 10 years, and $r = 0.5$. If q is doubled in size, what happens to the period of the cycle?

*6.3. Draw the state-space graph for a predator isocline with a carrying capacity and alternative prey, and a victim isocline with a carrying capacity and an Allee effect. Discuss predator–prey dynamics at the two intersection points.

*6.4. You are studying an insect-eating bird with a Type II functional response for which $k = 100$ prey/hour and $D = 5$.

a. What is the capture efficiency, α?

b. If the prey abundance is 75, what is the feeding rate (n/t)?

* Advanced problem

CHAPTER 7

Island Biogeography

Model Presentation and Predictions

THE SPECIES–AREA RELATIONSHIP

One of the few genuine "laws" in ecology is the **species–area relationship**: large islands support more species than small islands. The pattern holds for most assemblages of organisms, everything from vascular plants of the British Isles to reptiles and amphibians of the West Indies. The "islands" need not even be oceanic. Fish that live in lakes, mammals that occupy patches of forested mountaintops, and insects that visit thistle-heads all show a species–area relationship for their respective habitat islands. Because a national park or a nature reserve effectively is an island in a sea of disturbed habitat, studies of the species–area relationship may be relevant to the preservation of species in a fragmented landscape. This chapter explores in detail the relationship between area and the number of species in the community (species richness).

Figure 7.1a shows a typical species–area relationship for species of breeding land-birds on islands of the West Indies. The x axis shows the area of island, and the y axis shows the number of breeding land-bird species. You can see that the relationship is not linear: species number increases rapidly with area for small islands, but more slowly for large islands. For many oceanic islands, a rule of thumb (**Darlington's Rule**) is that a tenfold increase in island area results in a doubling of species number (Darlington 1957). Mathematically, the species–area relationship for many communities can be described by a simple power function:

$$S = cA^z$$

Equation 7.1

In this equation, S is the number of species, A is the island area, and z and c are fitted constants, which we will explain in a moment. If we take the logarithm (base 10) of each side of the equation, we have:

$$\log(S) = \log(c) + z\log(A)$$

Equation 7.2

The logarithmic transformation turns the species–area curve into a straight line when plotted on logarithmic axes. The constant $\log(c)$ is the intercept of that line, and the constant z is the slope of the line. Figure 7.1b shows the West Indian bird data replotted with both axes transformed to logarithms. The data conform fairly well to a straight line, suggesting a good fit to the power function.

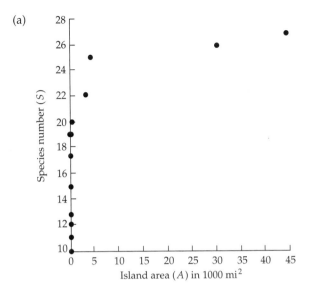

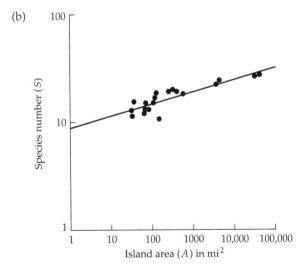

Figure 7.1 (a) Species–area relationship for breeding land-birds of the West Indies. Each point is a different island. Note that species number increases rapidly for small islands, but more slowly for large islands. (Data from Gotelli and Abele 1982.) (b) Logarithmic (base 10) transformation of the species–area relationship. The data in (a) have been plotted on a double-log plot. The best-fitting power function is shown by the straight line $\log(S) = 0.942 + 0.113 \log(A)$. Equivalently, the power function is $S = 8.759(A)^{0.113}$.

The area of an island is not the only factor that affects species richness. Figure 7.2 shows the effects of distance on species richness for birds of the Bismarck Islands in the tropical Pacific. New Guinea serves as the probable "source pool" for these islands, because all of the bird species found on the Bismarck Islands are a subset of the New Guinea avifauna. The x axis of this graph gives the distance from each island to New Guinea. The y axis shows the observed number of species divided by the number expected for a "near" island (< 500 km from New Guinea) of comparable area. You can see that relative species richness decreases with increasing distance from the source pool. In general, species richness is reduced for communities in small or isolated areas. In the following sections, we develop several models that attempt to explain the **area effect** (more species on large islands than on small islands) and the **distance effect** (more species on near islands than on far islands).

THE HABITAT DIVERSITY HYPOTHESIS

The most straightforward explanation for the species–area relationship is that large islands contain more habitat types than small islands. Therefore, species that are restricted to certain habitat types may occur only on large islands with those habitats. The species–area relationship for West Indian land-birds can be

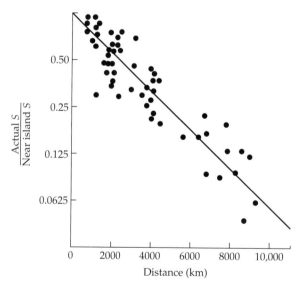

Figure 7.2 Distance effects for birds of the Bismarck Archipelago. The x axis gives the distance from each island to New Guinea, the presumed source pool. The y axis shows the observed species richness divided by the expected species richness for a "near" island (< 500 kilometers from New Guinea) of comparable size. (From Diamond 1972.)

explained, in part, by this phenomenon. The largest islands in the chain are the Greater Antilles (Puerto Rico, Cuba, Hispaniola, and Jamaica). These islands include many unique habitat types, such as extensive swampland (Cuba) and high-elevation pine forest (Hispaniola), that do not occur on any of the smaller islands. Habitat specialists such as the Zapata wren of Cuba (*Ferminia cerverai*) and the white-winged crossbill of Hispaniola (*Loxia leucoptera*) occur only in these particular habitats. Intermediate-sized islands, such as Guadeloupe and St. Lucia, are steep volcanic plugs that have fewer habitats and species than islands of the Greater Antilles. Some of the smaller islands, such as Antigua and Barbuda, are flat coral atolls. They are arid islands with structurally simple vegetation, and they support even fewer bird species.

Although habitat diversity can account for many species–area relationships, it is not always the correct explanation. For one thing, most species are not extreme habitat specialists, and their distribution may not always be limited by available habitat. In addition, there are many examples of species–area relationships in which there is little, if any, habitat variation. Within patches of identical habitat, species number is still greater on large islands than small, suggesting that other forces may be at work. In the next section, we develop the "equilibrium model" of island biogeography as an alternative hypothesis that accounts for the species–area relationship. Later in this chapter, we describe a third hypothesis, the passive sampling model, which can also explain the species–area relationship.

THE EQUILIBRIUM MODEL OF ISLAND BIOGEOGRAPHY

The **equilibrium model of island biogeography** was popularized by Robert H. MacArthur (1930–1972) and Edward O. Wilson (1929–). It is sometimes referred to as the "equilibrium model" or the "MacArthur–Wilson model." The model's basic premise is that the number of species occurring on an island represents a balance between recurrent *immigration* of new species onto the island, and recurrent *extinction* of resident species (MacArthur and Wilson 1963, 1967). When immigration and extinction rates are equal, the number of species is at an equilibrium. The concept is similar to the equilibrium N in a local population (Chapter 2), and to the equilibrium fraction of sites occupied by a metapopulation (Chapter 4).

The equilibrium model assumes there is a permanent mainland **source pool** of species that can potentially colonize an island. There are P species in the mainland pool, and we assume for now they are all similar to one another in colonization and extinction potential. We define the **immigration rate**, λ_s, as the number of new species colonizing the island per unit time. The **extinction rate**, μ_s, is the number of species already present on the island going extinct per unit time. The rate of change in species number on the island (dS/dt) is the difference between the immigration rate and the extinction rate.

Thus:

$$\frac{dS}{dt} = \lambda_s - \mu_s \qquad\qquad \text{Equation 7.3}$$

First, we will define the functions for λ_s and μ_s. Next, we will set Equation 7.3 equal to zero and solve for the equilibrium number of species. Finally, we will modify the extinction and immigration curves to account for the effects of area and isolation on species richness.

Figure 7.3 illustrates the immigration curve for the equilibrium model. The x axis of the graph shows the number of species present on the island. The y axis shows the rate of immigration. The **maximum immigration rate**, I, occurs when the island is empty. The immigration rate *decreases* as more species are added to the island. This is because as more species are present, fewer *new* species remain in the source pool as potential colonists. Finally, suppose that all of the species in the source pool are present on the island. By definition, there can be no further immigration, so the immigration curve crosses the x axis at the point $S = P$. Thus, the immigration curve is a decreasing straight line, with a maximum rate of I, and a minimum rate of zero, when $S = P$.

Remember that a straight line can be described by the equation $y = a + bx$, where a is the intercept and b is the slope. In this case, the intercept is I, and the slope (rise over run) is $-I/P$. Thus, the equation for the immigration rate is:

$$\lambda_s = I - \left(\frac{I}{P}\right)S \qquad\qquad \text{Expression 7.1}$$

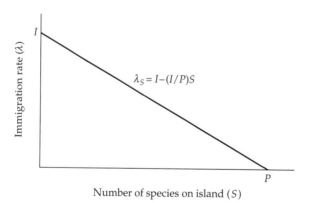

Figure 7.3 The immigration rate in the MacArthur–Wilson model. The immigration rate (number of species per unit time) decreases as more species are added to the island.

Now we turn to the extinction rate, μ_s. We expect μ_s to *increase* with increasing S: the more species present on the island, the greater the rate at which species disappear. This relationship occurs because each species has a constant probability of disappearance, so species disappear at a faster rate when there are more species present on the island. The **maximum extinction rate**, E, will occur when all the species in the source pool are present on the island ($S = P$). Conversely, if no species are present on the island ($S = 0$), the extinction rate must equal zero. Thus, the extinction curve is also a straight line with an intercept of zero, and a maximum rate of E, which occurs when $S = P$ (Figure 7.4):

$$\mu_s = \left(\frac{E}{P}\right)S \qquad \text{Expression 7.2}$$

Now that we have derived expressions for linear immigration and extinction rates, we can substitute these into Equation 7.3 to model the change in species richness on an island:

$$\frac{dS}{dt} = I - \left(\frac{I}{P}\right)S - \left(\frac{E}{P}\right)S \qquad \text{Expression 7.3}$$

The number of species on an island reaches an equilibrium when (dS/dt) equals zero. Setting Expression 7.3 equal to zero and solving for S yields:

$$S\left(\frac{I+E}{P}\right) = I \qquad \text{Expression 7.4}$$

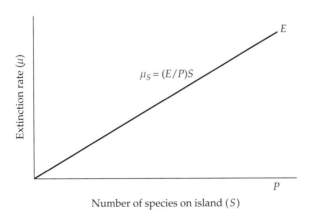

Figure 7.4 The extinction rate in the MacArthur–Wilson model. The extinction rate (number of species extinctions per unit time) increases as more species are added to the island.

The equilibrium number of species, $\hat{S}$, is thus:

$$\hat{S} = \frac{IP}{I+E}$$

Equation 7.4

The equilibrium depends on the size of the source pool (P) and the maximum immigration (I) and extinction (E) rates. Graphically, this equilibrium species number corresponds to the point on the x axis beneath the intersection of the immigration and extinction curves (Figure 7.5). At the intersection, the rate at which new species arrive is matched by the rate at which species present on the island go extinct.

This equilibrium point is stable. If we are below $\hat{S}$, we are to the left of the intersection point. In this region of the graph, the immigration rate exceeds the extinction rate, so species number increases. To the right of the intersection, extinctions exceed immigrations, so species number declines.

Equation 7.4 shows that species richness is increased by larger source pools and higher immigration rates, and decreased by higher extinction rates. Note the similarity between this equilibrium and the equilibrium in the island–mainland metapopulation model (Equation 4.4) we derived in Chapter 4. The intersection of the immigration and extinction curves also resembles the intersection of density-dependent birth and death rate curves in our derivation of the logistic growth equation (Figure 2.1) in Chapter 2.

Figure 7.5 also shows that the equilibrium is characterized by a **turnover rate**, which is measured on the y axis of the equilibrium graph. This turnover rate, T, is the number of species arriving (or disappearing) per unit time at equilibrium. T can be measured as either the extinction or the immigration rate, because these two are equal at equilibrium. Using some simple geometry, we see in Figure 7.5 that:

$$\frac{\hat{T}}{\hat{S}} = \frac{E}{P}$$

Expression 7.5

Therefore:

$$\hat{T}P = \hat{S}E$$

Expression 7.6

Rearranging and substituting Equation 7.4 for $\hat{S}$ gives:

$$\hat{T} = \frac{\left(\frac{IP}{I+E}\right)E}{P}$$

Expression 7.7

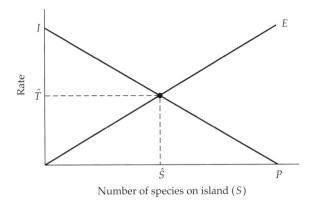

Figure 7.5 Equilibrium species number in the MacArthur–Wilson model. The intersection of the immigration and extinction curves determines the equilibrium number of species ($\hat{S}$) and the turnover rate ($\hat{T}$) at equilibrium.

$$\hat{T} = \frac{IE}{I+E}$$ Equation 7.5

Note that the turnover rate at equilibrium depends only on the maximum immigration and extinction rates (I and E), not on the size of the source pool (P). As you might expect, increasing either the maximum immigration or extinction rate increases the turnover at equilibrium.

This turnover of island populations at equilibrium is a key feature of the MacArthur–Wilson model. In contrast to many of the ecological models we have studied, the MacArthur–Wilson model does not predict stable populations. Instead, there is ongoing colonization and stochastic extinction of island populations. Species composition on the island is continually changing, although total species number remains relatively constant.

So far, we have constructed an equilibrium model of island species richness, but we still haven't explained the species–area effect. To do so, we must incorporate two additional assumptions about the demography of the colonizing species. The first assumption is that total population size for each species is proportional to island area. In other words, the *density* of populations (number of individuals per unit of area) is the same on islands of different size. The second assumption is that the probability of population extinction decreases with increasing population size. This assumption follows directly from the model of demographic stochasticity developed in Chapter 1. Because population sizes will be larger on big islands than on small islands, the extinction rates will correspondingly be lower on big islands.

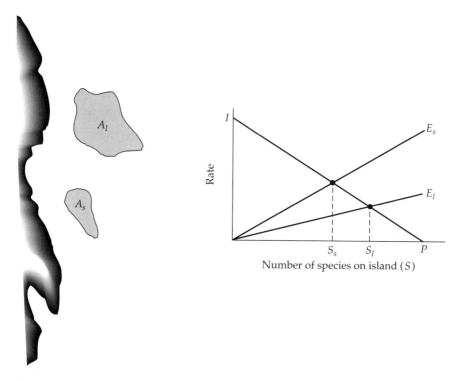

Figure 7.6 The area effect in the MacArthur–Wilson model. Smaller islands have smaller population sizes, which increases the extinction rate and leads to a lower species equilibrium. E_s is the maximum extinction rate for small islands; E_l is the maximum extinction rate for large islands.

Suppose we have a large island (A_l) and a small island (A_s) that differ only in area, but are identical in habitat diversity and distance from the source pool (Figure 7.6). Because both islands are equidistant from the mainland and colonized by the same source pool of P species, they have the same immigration curve. However, maximum extinction rates on the large island (E_l) are lower than on the small island (E_s) because population sizes are greater on the large island. Because of this area effect, the equilibrium number of species is greater on the large island, with a lower rate of turnover.

We can also account for the distance effect by modifying the immigration curves for near and distant islands. Suppose two islands have identical areas and habitats, but differ in their distance from the source pool (Figure 7.7). Because the areas are equal, the two islands have the same extinction curve. But the maximum immigration rate will be higher on the near island (I_n) than

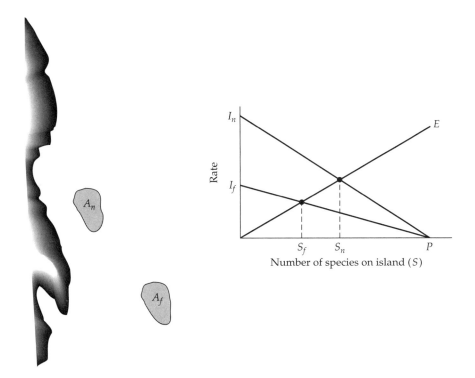

Figure 7.7 The distance effect in the MacArthur–Wilson model. Islands that are distant or isolated from the source pool have a reduced immigration rate, leading to a lower species equilibrium. I_n is the maximum immigration rate for near islands; I_f is the maximum immigration rate for far islands.

on the far island (I_f). Consequently, the near island will have more species at equilibrium than the far island. The near island will also be characterized by greater turnover than the far island.

Thus, island species richness in the MacArthur–Wilson model is uniquely determined by the geometry of an island—its area determines the extinction rate, and its distance or isolation determines its immigration rate. The intersection of these two curves controls the equilibrium number of species and the turnover rate.

Model Assumptions

Although the equilibrium model predicts patterns of species richness, its underlying assumptions are at the population level. These assumptions are:

✔ **An island potentially can be colonized by a set of P source pool species that have similar colonization and extinction rates.** This assumption implies that the species in the source pool and on the island are not undergoing any evolutionary change that might alter colonization or extinction rates. Thus, like most ecological models, the equilibrium model does not incorporate evolutionary mechanisms or historical constraints that might influence species richness.

✔ **The probability of colonization is inversely proportional to isolation or distance from the source pool.** Isolated islands have shallower immigration curves than non-isolated islands. All other things being equal, this lowers the equilibrium number of species (see Problem 7.2).

✔ **The population size of a given species is proportional to the area of the island.** In other words, the density of each population (number of individuals/area) is constant throughout the archipelago. Alternative models (Schoener 1976) assume that competitive interactions are important, so that both island area and species richness influence population size.

✔ **The probability of a population becoming extinct is inversely proportional to its size.** Although the equilibrium model does not explicitly forecast population sizes, this assumption incorporates the idea that demographic stochasticity (see Chapter 1) increases the risk of extinction at small population sizes. This assumption and the previous one ensure that extinction curves are steeper for small islands than they are for large islands, leading to a species-area curve.

✔ **Colonization and extinction of local populations is independent of species composition on the island.** In contrast to classic models of competition (Chapter 5) and predation (Chapter 6), the equilibrium model assumes that the presence of one species does not affect the colonization or extinction of another. If extinctions are independent of species composition, the island community is "non-interactive." If colonizations are independent, the island community is not undergoing any successional change, because the particular order of arrival and departure of species is not important.

Model Variations

NONLINEAR IMMIGRATION AND EXTINCTION CURVES

The linear immigration curve implies that all species have identical potential for dispersal and colonization of islands. But suppose that some species are

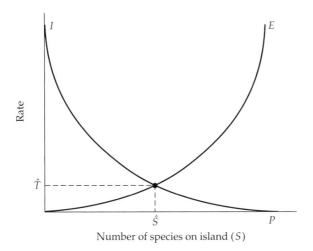

Number of species on island (S)

Figure 7.8 Nonlinear immigration and extinction curves in the MacArthur–Wilson model. These curves may reflect the influence of species interaction on the extinction rate and differential colonization ability on the immigration rate. The qualitative predictions of this nonlinear model are similar to those of the linear model described in the text (Figure 7.5).

much better at dispersal and colonization than others. These species would be among the first to colonize an empty island, whereas the poor dispersers would arrive later in the colonization sequence. With differential dispersal, the immigration curve would be exponential, with a steep decline initially and a slower rate of decrease as later species are added (Figure 7.8).

Similarly, the linear extinction curve implies that species extinctions are independent of one another. It might be more realistic to assume that competition increases the extinction rate when more species are present. In this case, the extinction curve would increase exponentially with S (Figure 7.8). In textbooks, the MacArthur–Wilson model is usually presented with these nonlinear immigration and extinction curves. Fortunately, the basic predictions of the equilibrium model remain the same, whether the linear or nonlinear rate curves are used.

AREA AND DISTANCE EFFECTS

Both area and distance affect extinction and immigration in the MacArthur–Wilson model. But the basic model describes only two mechanisms: the effect of area on extinction, and the effect of distance on immigration (Figure 7.9). In the next two sections, we briefly explore the effect of distance on extinction (the "rescue effect"), and the effect of island area on immi-

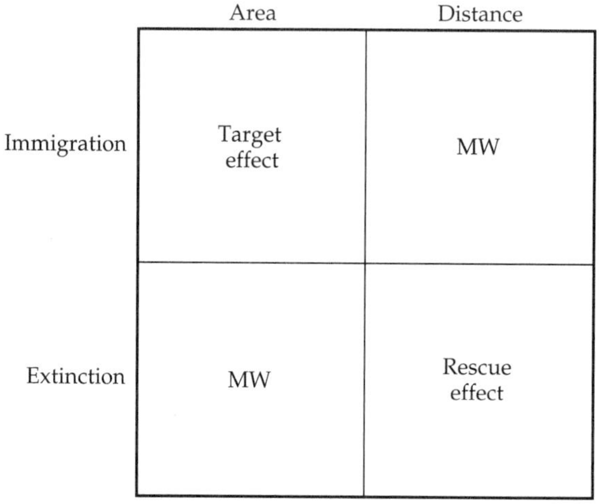

	Area	Distance
Immigration	Target effect	MW
Extinction	MW	Rescue effect

Figure 7.9 Area and distance effects in the MacArthur–Wilson model. The basic model (MW) considers the effects of area on the extinction rate and distance on the immigration rate. The model can be extended to incorporate the effects of distance on the extinction rate (rescue effect), and the effects of area on the immigration rate (target effect).

gration (the "target effect"). These modifications incorporate more biological realism, but they also complicate the predictions of the simple MacArthur–Wilson model. We then develop a "passive sampling" model that may also account for the species–area relationship without invoking habitat specialization or a species equilibrium.

THE RESCUE EFFECT

The MacArthur–Wilson model assumes that the only effect of distance or isolation is on the immigration rate. However, as we saw in Chapter 4, isolation can also affect the probability of extinction. In the metapopulation models of Chapter 4, we defined the rescue effect as the reduction in the probability of local extinction as the frequency of occupied patches increases. For the island model, we can define the rescue effect as the reduction in the species extinction rate for near versus far islands (Brown and Kodric-Brown 1977). Figure 7.10 illustrates the change in the immigration and extinction curves when a rescue effect is present. The basic prediction that large islands have more species than small ones remains unchanged with the rescue effect. However, the original MacArthur–Wilson model predicted *less* turnover on more isolated islands because they received fewer immigrants. In contrast, the rescue effect may generate *greater* turnover on more isolated islands because of the increase in the extinction rate.

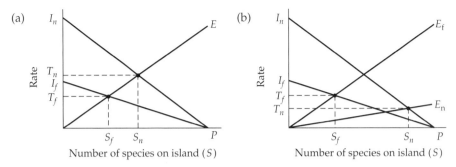

Figure 7.10 The rescue effect is the reduction in the extinction rate of near islands versus distant ones. Whereas the simple MacArthur–Wilson model predicts higher turnover on near islands (a), the rescue effect may increase turnover on more distant islands (b). T_n is the turnover rate on the near island; T_f is the turnover rate on the far island.

THE TARGET EFFECT

The MacArthur–Wilson model considers only the effects of area on the extinction rate. However, island area might affect the immigration rate as well. To the extent that islands functions as targets that intercept colonizing individuals, large islands may have higher immigration rates than small islands (Lomolino 1990). We can incorporate this **target effect** by assuming that the immigration rate is higher on large islands than on small. As in our analysis of the rescue effect, this change does not alter the pattern of species richness on large and small islands. If the target effect is strong enough, the model still predicts a species–area relationship, but turnover is now greater on large islands than on small (Figure 7.11).

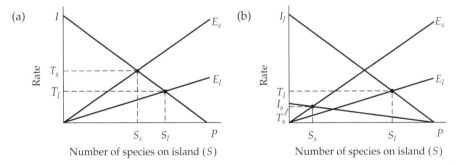

Figure 7.11 The target effect is the increase in the immigration rate on large islands versus small ones. Whereas the simple MacArthur–Wilson model predicts higher turnover on small islands (a), the target effect may increase turnover on large islands (b). T_s is the turnover rate on the small island; T_l is the turnover rate on the large island.

THE PASSIVE SAMPLING MODEL

The nonlinear rate curves, rescue effect, and target effect are straightforward variations of the MacArthur–Wilson equilibrium model. All of these variations still describe species richness as a balance between ongoing immigration and ongoing extinction. But might there not be a simpler explanation for the species–area relationship? Suppose that islands function as passive "targets" that randomly accumulate individuals. Even in the absence of equilibrium turnover or habitat effects, we would still expect large islands to accumulate more species, by chance alone.

A useful analogy is to think of the islands as a set of targets. The area of each island is equivalent to the area of each target. Each individual organism is a dart, which is tossed randomly at the set of targets. The different species are represented by different colors of darts. Suppose we toss a handful of these darts at the targets. By chance, we expect the larger targets to accumulate more darts, and hence more colors, than the smaller targets. Similarly, if individuals colonize islands randomly, large islands should accumulate more individuals and species than small islands.

We can use some simple principles of probability theory to develop this **passive sampling model** (Coleman et al. 1982). First, assume we have a set of k islands. We will use the counter i to denote the ith island. The area of the ith island in the list is denoted as a_i. For example, if the fifth island on our list has an area of 100 square miles, $a_5 = 100$. Similarly, we assume a set of s species. We will use the counter j to denote the jth species. The total abundance of species j (summed across all islands) is n_j. If there are a total of 300 individuals of the sixth species that occur in the archipelago, $n_6 = 300$.

Let A equal the summed area of all the islands:

$$A = \sum_{i=1}^{k} a_i \qquad \text{Expression 7.8}$$

Next, define x_i as the **relative area** of the ith island:

$$x_i = \frac{a_i}{A} \qquad \text{Expression 7.9}$$

Note that these proportional areas sum to 1.0:

$$\sum_{i=1}^{k} x_i = 1.0 \qquad \text{Expression 7.10}$$

x_i can also be interpreted as the *probability* that a randomly placed individual will intercept an island of area a_i. Therefore, the probability that a single individual will *not* reach a particular island is:

$$P(1 \text{ miss}) = 1 - x_i \qquad \text{Expression 7.11}$$

For species j, the probability that *all* n_j individuals miss the island is:

$$P\left(n_j \text{ misses}\right) = \left(1 - x_i\right)^{n_j} \qquad \text{Expression 7.12}$$

Expression 7.12 gives the probability that none of the n_j individuals of species j land on the island. Therefore, the probability that *at least* one individual of species j is present on the island is:

$$P(\text{species } j \text{ occurs on island } i) = 1 - \left(1 - x_i\right)^{n_j} \qquad \text{Expression 7.13}$$

Finally, if we sum these probabilities across all species, we obtain the expected species richness on island i [$E(S_i)$]:

$$E(S_i) = \sum_{j=1}^{S} \left[1 - \left(1 - x_i\right)^{n_j} \right] \qquad \text{Equation 7.6}$$

Why should the expected species richness equal the sum of the probabilities of species occurrence? Suppose that the probability of occurrence of each species was 0.5. Intuitively, you would expect to find about half of the species in the archipelago occurring on the island. This expectation has a variance associated with it (Coleman et al. 1982), but the derivation is beyond the scope of this primer.

Like the MacArthur–Wilson model, the passive sampling model predicts more species on large islands than on small. However, the MacArthur–Wilson model predicts recurrent extinction and turnover of island populations, whereas the passive sampling model does not invoke turnover. Instead, the passive sampling model predicts that an abundant species will have a greater chance of occurring on an island than a sparse species. In fact, if an island is extremely small, rare species may be unlikely to ever occur there. Thus, the passive sampling model gives us more predictive power about species composition than does the MacArthur–Wilson model. The passive sampling model does not explicitly account for the distance effect, although we could extend the theory by modifying the relative target area as a function of distance from the source pool.

Empirical Examples

INSECTS OF MANGROVE ISLANDS

The most famous test of the equilibrium model was by Edward O. Wilson and his student, Daniel Simberloff. These authors studied the insects that colonized small mangrove "islands" in the Florida Keys (Wilson and Simberloff 1969, Simberloff and Wilson 1969). Each island consisted of one to several red mangrove trees (*Rhizophora mangle*) that grow in shallow seawater. The total arthropod source pool for these islands was approximately 250 species, and each island supported 20 to 50 species. There are thousands of these mangrove islands in the Florida Keys that differ in their area and distance from colonization sources.

Simberloff and Wilson chose six islands for experimental manipulation and carefully censused them at the start of the experiment. The islands were then covered with canvas and all the resident arthropods were killed with methyl bromide, an insecticide. Over the next year, the authors repeatedly censused the islands and recorded the presence of different insect species during recolonization. The basic predictions of the equilibrium model were confirmed: after 250 days, species number on most islands had returned to approximately the same level as before defaunation (Figure 7.12). Large, near islands accumulated more species than small, distant islands.

Perhaps more important, the census data revealed considerable turnover in species composition, which is the essential prediction of the MacArthur–Wilson model. Figure 7.13 shows part of the recolonization records for one of the six experimental islands. Although species number returned to an approximate equilibrium, species identity changed considerably from one census to the next, with an estimated turnover rate of 0.67 species per day.

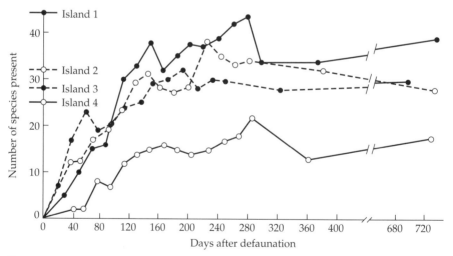

Figure 7.12 Insect recolonization of four defaunated mangrove islands. The *y* axis indicates the predefaunation species richness of each island. Most of the islands reached an equilibrium species number after 250 days that was approximately the same as the initial richness. (From Simberloff and Wilson 1969.)

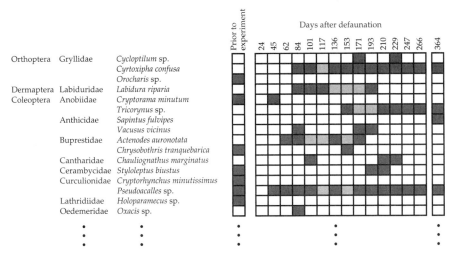

Figure 7.13 Colonization and extinction records for a single mangrove island. Each row is a species and each column is a census date. Records are shown for 16 of the 90 arthropod species that colonized the island. Open squares indicate species absence; darker squares indicate species presence. Lightly shaded squares indicate a species was not seen, but its presence was inferred from other evidence. Note the substantial turnover and change in species composition from one census to the next. (From Simberloff and Wilson 1969.)

However, Simberloff (1976) re-analyzed the mangrove data and questioned whether there was actually this much turnover. He pointed out that it is important to distinguish between local extinctions of relatively isolated breeding populations, and transient, short-term movements of individuals between islands. From the original census records, he eliminated those populations that were represented by only one or two individuals because they were unlikely to represent breeding populations. He also eliminated those populations that disappeared before they would have had time to reproduce. The corrected estimate of turnover was only 1.5 extinctions per year! Simberloff (1976) concluded that a test of the equilibrium theory required careful definitions of what constituted a true "colonization," and that much of the observed turnover in the mangrove insect community was among transient species.

BREEDING BIRDS OF EASTERN WOOD

Although the immigration and extinction curves (Figure 7.5) are the heart of the equilibrium model, they have rarely been measured in the field. An interesting exception is a long-term study of bird populations in a small plot within an oak forest (Williamson 1981). From 1947 through 1975, a team of ornithologists annually censused Eastern Wood, a 16-hectare plot of oak woodland in Surrey, England. The record of extinctions and colonizations can be plotted as a function of the number of species present each year, which varied from 27 to 36. The immigration curve matched the basic prediction of the equilibrium model, declining from an estimated maximum of 16 species per year to a value of zero at 40 resident species. This is somewhat less than the source pool estimate of 44 species. As predicted by the MacArthur–Wilson model, the extinction rate increased with S, although there was so much scatter in the data the trend was not statistically significant (Figure 7.14).

As in Simberloff's (1976) analysis, a detailed consideration of breeding status complicates the picture of species equilibrium. A core group of 14 species bred in the wood every year. A second group of 19 species did not establish substantial populations. These included species whose breeding status had not been confirmed (6), species that were represented by only 1 or 2 pairs in the plot (9), and species whose territories were larger than the area of the plot (4). The remaining 11 species were casual breeders that underwent frequent extinction.

The fit of the MacArthur–Wilson model to these data is somewhat ambiguous. On the one hand, the existence of ecological turnover and the qualitative appearance of the immigration and extinction curves (Figure 7.14) match-

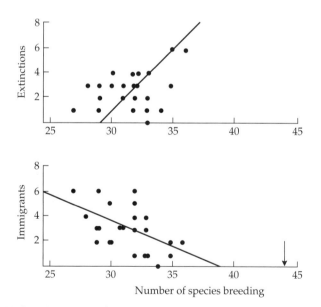

Figure 7.14 Immigration and extinction rates (species per year) for breeding landbirds of Eastern Wood. The immigration curve declined significantly with increasing species richness, but the extinction curve showed only a weak positive association with species richness. The arrow indicates 44 species, the source pool estimate. Compare to Figures 7.3 and 7.4. (From Williamson 1981.)

es the basic assumptions of the equilibrium model. On the other hand, the occurrence of 14 core species that bred every year in the woods could not have been predicted by the equilibrium model. This type of community structure may be fairly typical—one set of species with stable, persistent populations, and a second set of species with transient populations, which frequently go extinct and re-colonize. The models we have developed in Chapters 1, 2, 3, 5, and 6 may be more appropriate for the persistent populations, whereas the models in this chapter and in Chapter 4 may be more appropriate for transient populations.

BREEDING BIRDS OF THE PYMATUNING LAKE ISLANDS

Although the passive sampling model was introduced over 70 years ago (Arrhenius 1921), it has only received widespread attention since the early 1980s. Coleman et al. (1982) developed the mathematical predictions of the passive sampling model, and tested it with data on island breeding birds. A number of islands in Pymatuning Lake on the Ohio–Pennsylvania border were thoroughly censused for nests and bird territories. These islands were originally hilltops before a reservoir was created in 1932. The islands retain their deciduous forest vegetation, and the archipelago supports a pool of approximately 36 breeding land-bird species.

Coleman et al. (1982) knew the area of each island, and they were able to estimate the abundance of each species on the islands. They used these data to predict island species richness with the passive sampling model. In Figure 7.15, the solid line shows the predicted species richness and a confidence interval based on the passive sampling model. Species richness on most islands matched this prediction fairly well. In fact, the passive sampling model did a better job of predicting island species richness than did the power function.

One drawback of the passive sampling model is that it requires estimates of the abundance of all species on all islands, and these may be difficult to obtain. A second drawback is that the target analogy is conceptually simple, but biologically not very realistic. Many factors other than island area affect individual colonization, including weather and current patterns, seasonal migration, food resources, and the presence of other species of predators and competitors.

In conclusion, the species–area relationship represents one of the few general patterns in ecology, but its causes remain elusive. The habitat diversity hypothesis, the MacArthur–Wilson equilibrium model, and the passive sampling model are not mutually exclusive explanations—each may contribute to the species–area relationship. Additional data on habitat diversity, population turnover, and source pool structure are needed to gauge their relative contributions to the species–area relationship.

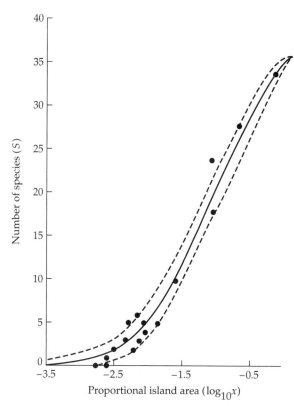

Figure 7.15 Observed and expected species richness for breeding land-birds on islands in Pymatuning Lake. The x axis gives the logarithm of the fractional area of each island. The solid line is the expected species richness and the dashed lines show the confidence interval from the passive sampling model. Each circle is the observed species richness for each island. Note the good match between the observed data and the model predictions. (From Coleman et al. 1982.)

Problems

7.1. For the West Indian land-bird data in Figure 7.1, the best-fitting power function has the constants $c = 8.759$ and $z = 0.113$. The island of Grenada has an area of 120 square miles, and supports 17 land-bird species.

   a. What is the predicted number of species from the power function?

   b. Suppose that half of the island's area disappears in a volcanic eruption. From the power function, how many species would be expected to remain?

7.2. Your colleague returns from the South Pacific with data on island lizards. "Look," she says, "my data show that there are more species of lizards on small islands than large islands. This disproves the MacArthur–Wilson equilibrium model!" Using an appropriate set of immigration and extinction curves, show how more species could occur on a small island (A_2) than on a large island (A_1) in the MacArthur–Wilson model.

7.3. Suppose that an island in MacArthur–Wilson equilibrium supports 75 species, out of a source pool of 100 species. The maximum extinction rate (E) is 10 extinctions per year. Calculate the maximum immigration rate (I). If I is doubled, what is the new species equilibrium and new turnover rate?

*7.4. Here are some hypothetical data on the abundances of six species of cactus on four small desert islands:

	Island 1 (110 ha)	Island 2 (100 ha)	Island 3 (10 ha)	Island 4 (5 ha)
Cactus A	3	0	0	0
Cactus B	1	0	0	0
Cactus C	4	2	3	1
Cactus D	2	0	2	2
Cactus E	1	0	1	0
Cactus F	1	0	0	3

Calculate the expected number of species on each island for the passive sampling model. How close are the expected values to the observed number of species?

* Advanced problem

CHAPTER 8

Succession

Model Presentation and Predictions

When ecologists study communities, they often make comparisons across space. We might compare the ant species of a bog and a nearby forest, or the wildflowers of two alpine meadows on a mountainside. These "snapshot" comparisons form the basic stuff of community ecology—we are interested both in describing how communities differ from one place to another and in understanding the processes that lead to those differences (Wiens 1989).

However, communities do not spontaneously appear in their current state, and they do not stay the same through time. Instead, communities are built up gradually through colonization, and community structure changes through time (Huston 1994). This chapter focuses on the details of those changes.

By studying the temporal dynamics of communities we may be able to understand the varieties of communities that we encounter in different places. Rather than visiting, say, 100 forest sites during the course of a busy field season, we would prefer to sit beneath a single tree and watch the scenery around us change over the course of 100 years. This sort of "trajectory" experiment would reveal the dynamics of the community and the mechanisms that lead to change through time (Diamond 1986).

Of course, we cannot carry out trajectory experiments on these time scales. Paleoecologists have used fossil series and pollen profiles to successfully reconstruct communities of the past (e.g., Spear et al. 1994). However, these studies usually give a coarse picture of community change, often over time scales of thousands of years. In this chapter, we will describe some methods for studying community change that can be applied over shorter time intervals, such as years and decades.

THREE VERBAL MODELS OF SUCCESSION

Succession—broadly defined—is the change in community structure through time. We will briefly describe some of the ideas and concepts that have been important in the study of succession, and then develop a simple mathematical model to describe those changes.

The process of succession begins with an "empty" community that contains no species. **Primary succession** occurs when a new substrate is formed and colonized, as when a volcano erupts or a glacier retreats. More common is **secondary succession**, in which a previously established community is removed by a disturbance. These disturbances "re-set the clock" of succession and are a universal feature of communities. Everything from treefalls and fires to ice scour and hurricanes constitute natural disturbances that can initiate the process of succession. The most widespread forms of disturbance now come from human activities, including clearcutting and burning of forests, agricultural plantings, urbanization, and harvesting of natural and managed populations. When humans abandon these activities, successional change can begin again, though perhaps no longer with the same trajectory.

How do species reenter a community following a disturbance? During secondary succession, many elements of the previous community may re-establish themselves. Dormant seeds, resistant egg or larval stages, and regenerating adults that were damaged but not killed may re-appear in a disturbed patch. But the most common source of colonists will be dispersing individuals, both juveniles and adults. These dispersers originate from nearby patches that were not disturbed.

Early on, ecologists made two interesting observations about colonization following a disturbance. The first observation was that the species that showed up immediately following a disturbance were often very different from the species that would show up later in time. These **pioneer species** have life history traits that allow them to thrive in the harsh physical conditions of a newly disturbed patch. These traits include high fecundity and dispersal potential, rapid population growth rate, and low competitive ability—in short, many of the same life history traits that characterize r-selected species (see Chapter 3). As we shall see, these pioneering species do not persist indefinitely, and are eventually replaced by other species.

The second observation was that communities that are disturbed in different ways and look different to begin with may become more similar through time. For example, forest patches that are damaged by windstorms, cleared for agriculture, or selectively logged may—though not with certainty—converge to a similar structure of secondary forest growth a century after the last disturbance.

These observations—the presence of pioneer species and the convergence of communities following disturbance—suggested that changes in species composition during succession were deterministic, not random. Several models of succession have been proposed, and the most famous (and oldest) is the **facilitation model**. In the facilitation model, a newly disturbed patch is colonized by a set of pioneer species, which are the only ones that can survive the harsh physical conditions following a disturbance (Clements 1904). Through their growth and presence, pioneer species alter the physical environment in the patch. In terrestrial succession, pioneering plant species often stabilize soil movement, shade the soil surface, and add nutrients to the soil profile when individuals die and decay. These alterations facilitate, or "pave the way," for the next group of species to invade while simultaneously making the environment less suitable for the original set of pioneer species. For example, tree seedlings may only become established after grasses have initially invaded a patch, but the grasses will eventually die out beneath the shade of the trees.

Because of changing environmental conditions and competitive interactions, pioneer species eventually disappear and are replaced by a second set of species. These species also alter the environment and pave the way for additional groups of species to enter in order. In the classic facilitation model, the endpoint of this process is a so-called **climax community**, which is not replaced by any other group of species. The climax community is invasion-resistant and self-replacing, at least until another disturbance re-sets the system (Clements 1936). The key prediction of the facilitation model is that succession proceeds in a predictable sequence of community replacements: Community A (comprised of pioneer species) is replaced by community B, community B is then replaced by community C, and so on until the climax community is reached. The sequence must occur in order because facilitative changes in the environment are necessary before the next group of species can colonize.

Although the evidence for pioneer species during the early stages of colonization is strong, the picture is not so clear at the endpoint of the sequence. Most ecologists have abandoned the idea of a simple climax community. Initial colonization events, influences of habitat type and long-term environmental change, and historical and evolutionary factors lead to idiosyncratic endpoints in species succession (Facelli and Pickett 1990). Other ecologists have pointed out that facilitation is not the only game in town (Connell and Slatyer 1977; Huston 1994). For example, marine communities that develop on pier pilings or rock walls (see "Population Dynamics of Subtidal Ascidians" in Chapter 2) are often dominated by a single species, such as a barnacle or tunicate, that initially colonizes and then monopolizes the available space (Sutherland 1974).

These invaders are not pioneer species in the sense of the classic facilitation model. Instead, they manage to hang on to their initial "land grab," and may not be displaced until another disturbance re-sets the system (Law and Morton 1993). In this **inhibition model**, initial colonists suppress the arrival of later species, whereas in the facilitation model, pioneer species promote the entry of latecomers.

Finally, we should consider the proposition that initial colonists neither inhibit nor facilitate the arrival of later species. This **tolerance model** of succession can be viewed as a kind of simple null hypothesis, in which biotic interactions and alterations of the environment do not have a great influence on succession. However, an orderly successional sequence may still arise in the tolerance model because of differences in species life histories and colonization potential (Connell and Slatyer 1977). There are more complex ways to categorize successional models (Noble 1981), but these three models incorporate the most important mechanisms.

MATRIX MODELS OF SUCCESSION

The facilitation, inhibition, and tolerance models are *verbal* descriptions of how succession might proceed. We want to produce a *mathematical* model that describes those processes more precisely and generates predictions that can be tested in the real world. In order to do this, we will develop a matrix model of ecological succession that describes, in general terms, the ways in which communities change from one "state" to another through time. Then we will set the parameters of that model to describe the processes of facilitation, inhibition, and tolerance.

The matrix model that we will use is a simple but powerful method for describing changes in populations and communities (Horn 1975; Usher 1979). In fact, you have already been introduced to this **Markov model*** in Chapter 3, when we discussed age-structured growth. Both the simple Leslie matrix model and the more advanced stage-based projection model use exactly the same machinery of matrix multiplication that we will use here to model succession. (If you haven't done so already, you should go back and read Chapter 3 before proceeding.)

SETTING THE STAGES

To begin with, we need to define a number of mutually exclusive stages that represent different, discrete communities. These stages may represent entire sets of species (e.g., "algal mat," "encrusting sponges"), or they could even represent individuals or stands of a single species (e.g., "red maple," "shagbark hickory"). Some decisions must be made in order to organize and classify natural communities into a handful of discrete stages that are recognizable and useful to us as ecologists. Whether these represent true "natural" units of succession is an entirely different question, and it is not one that can be answered by the model itself.

Incidentally, choosing the stages for the model implicitly sets the spatial scale of the patch. For example, if the stages represent individual species, then a single patch must not be so large as to hold more than one species at a time. In addition to the stages that represent different community types, the model will probably also include an "open space" stage, which represents the community state immediately after a disturbance. Whatever stages are ultimately chosen, they must be *mutually exclusive* and *all-encompassing*. In other words, at any given time, a community can be readily classified as one, and

---

*Andrei Andreyevich Markov (1856–1922) was a Russian mathematician who pioneered the study of sequences of random variables in which the future variable is determined by the present variable but is independent of the way in which the present state is reached.

only one, of the specified stages (mutually exclusive), and the stages must include all of the possibilities for a patch (all-encompassing).

SPECIFYING THE TIME STEP

Once the stages are established, the investigator must specify the time step of the model. Unlike the differential equations we have mostly used in previous chapters, the stage-based successional model occurs in discrete time, and the time steps must be specified. Typically, a successional model might have a time step of a year or a decade. However, short-lived or seasonal assemblages, such as ephemeral algae or insects that colonize carrion, might be modeled with a time step of weeks or even days.

CONSTRUCTING THE STAGE VECTOR

Suppose we have a set of n possible stages for a patch. Imagine that the landscape consists of a large number of such patches. We can then create a **stage vector** that tells us the number of patches in each of the stages. We will use a boldfaced **s** as a shortcut to indicate such a vector.

For example suppose we have decided to model successional change among four patch types: open space, grassland, shrub, and forest. If we went out and censused a set of 500 patches, we might have the following values:

$$\mathbf{s}(t) = [250, 100, 80, 70] \qquad \text{Expression 8.1}$$

The expression in parentheses (t) tells us this is the stage vector at time t. Thus, 250 patches are open space patches, 100 are grassland patches, 80 are shrub patches, and 70 are forest patches. The entries in the patch vector must be non-negative real numbers. Zeroes are possible and indicate that one of the stages is not currently represented by any of the patches. Our model will show us how these numbers change through time and predict the distribution of patch types at equilibrium.

CONSTRUCTING THE TRANSITION MATRIX

In order to get there, we now introduce the centerpiece of this model: the **transition matrix**. This matrix is equivalent to the Leslie matrix used to forecast population growth in Chapter 3. However, there are some important differences in how we interpret the transitions, and in the possible values that can be entered in this matrix.

If there are n stages in our model, the transition matrix **A** will be square, with n rows and n columns. The labels for the rows and columns of this matrix are just the patch names themselves. Each column of the matrix represents the patch state at the current time (t), and each row represents the patch state at the next time step ($t + 1$). The entries in the matrix are the transition probabilities for change from the current state (column) to the next state (row). (We will explain the interpretation of these transitions in detail in just a bit.) Thus, in our four-stage example, we have a transition matrix with 4 rows, 4 columns, and 16 entries. Here is a typical matrix that we could have:

		Stage at time t			
		Open	*Grassland*	*Shrub*	*Forest*
Stage at time t + 1	*Open*	0.65	0.23	0.25	0.40
	Grassland	0.15	0.70	0.25	0.10
	Shrub	0.00	0.07	0.25	0.15
	Forest	0.20	0.00	0.25	0.35

Expression 8.2

The entries in the matrix tell us the probability p_{ij} of moving from stage j in time step t to stage i in time step $t + 1$. For example, the probability of moving from grassland (column) to the open state (row) is 0.23, and the probability of moving from forest to grassland is 0.10.

The stage transitions are not necessarily symmetric. For example, although the probability of moving from grassland to the open state is 0.23, the probability of moving from the open state to grassland is 0.15. The diagonals of the matrix indicate the probability that a patch remains in its current state, and does not change to any of the other states during a single time step.

This example illustrates some general properties of transition matrices for succession models. First, notice that all of the entries are positive numbers that are between 0.0 and 1.0. That makes sense, because these matrix elements represent *probabilities* of change, and probabilities cannot be larger than 1.0 or smaller than 0.0. Second, notice that all of the values in any column of the matrix add up to exactly 1.0. Why should this be so? Earlier we said that the stages in this model were mutually exclusive and all-encompassing. Because these are the only possible states for a patch, the probabilities must sum to 1.0 for all of the events that can occur once a patch is in a particular state.

LOOP DIAGRAMS

The transition matrix can be represented graphically as a loop diagram. To construct a loop diagram, draw a circle to represent each stage in the model. Use a one-headed arrow to connect two stages, and write the value of the transition above the arrow. Do not draw an arrow for any transitions that have a value of 0.0. The transition matrix in Expression 8.2 is diagrammed in Figure 8.1.

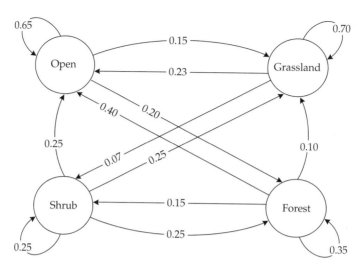

Figure 8.1 Loop diagram for the transition matrix in Expression 8.2.

PROJECTING COMMUNITY CHANGE

The transition matrix summarizes all of the information on how patches change from one state to another. The matrix is a set of probabilistic "rules" that determine the patterns of succession that will occur from any possible starting point. The next step is to apply those rules to the stage vector **s**. Specifically, if we have the vector **s** at time t, we can use the transition matrix to determine the number of patches in each state at the next time step ($t + 1$). In compact matrix notation, this amounts to multiplying the transition matrix **A** by the stage vector **s** at time t to produce the stage vector **s** at time $t + 1$:

$$\mathbf{s}(t + 1) = \mathbf{As}(t) \qquad \text{Equation 8.1}$$

In our example, we started with 100 patches in the grassland state at time t. How many patches will be grassland at time $t + 1$? The answer comes from multiplying each entry in the appropriate row of the matrix by the corresponding number of patches in the stage vector. There are four ways that grassland patches can appear at time $t + 1$: There can be transitions from patches that were previously in the open, shrub, or forest state, and there can be grassland patches that remain grassland patches. From our matrix, we have:

Grassland patches ($t + 1$) = (0.15)(250) + (0.70)(100) + (0.25)(80) + (0.10)(70)
= 134.5

Expression 8.3

The first term in this multiplication says that, of the 250 patches in the open state, 15% will transform to grassland patches [(0.15)(250)]. We add to this the contributions coming from shrubland patches [(0.25)(80)] and forest patches [(0.10)(70)]. Finally, we include the 70% of grassland patches that remain as grassland during this transition [(0.70)(100)]. Thus, we started with 100 grassland patches at time t, and we end up with 134.5 grassland patches at time $(t + 1)$.

The other transitions are calculated as:

Disturbed patches $(t + 1)$ = (0.65)(250) + (0.23)(100) + (0.25)(80) + (0.40)(70)
= 233.5

Shrub patches $(t + 1)$ = (0.00)(250) + (0.07)(100) + (0.25)(80) + (0.15)(70)
= 37.5

Forest patches $(t + 1)$ = (0.20)(250) + (0.00)(100) + (0.25)(80) + (0.35)(70)
= 94.5

$$\text{Expression 8.4}$$

Thus, we started with the vector:

$$\mathbf{s}(0) = [250\ 100\ 80\ 70] \qquad \text{Expression 8.5}$$

and, after one time step, ended up with the vector:

$$\mathbf{s}(1) = [134.5\ 233.5\ 37.5\ 94.5] \qquad \text{Expression 8.6}$$

Although the patch states have changed from one time step to the next, note that the total number of patches (500) remains the same. Now that we have the stage distribution in time $t + 1$, we can use that as the model input to project the stages in the next time step:

$$\mathbf{s}(t + 2) = \mathbf{As}(t + 1) \qquad \text{Equation 8.2}$$

At each step of the model, we multiply the transition matrix $\mathbf{A}$ by the current stage vector to generate the next set of states. As we will see, although our numbers change initially, they eventually stop changing.

DETERMINING THE EQUILIBRIUM

Even though the content of the stage vector keeps changing, remember that the transition matrix $\mathbf{A}$ remains the same through this process, and is used to multiply the stage vector at each step. If we continue to multiply the current stage vector by the projection matrix, we find that the stage vector quickly reaches the following equilibrium state:

$$\mathbf{s}(t) = [223.03 \ 164.70 \ 31.52 \ 80.75] \qquad \text{Expression 8.7}$$

Once this distribution is reached, there will be no further change in these numbers, no matter how many times we carry out the matrix multiplication. Even more interesting, this same equilibrium vector is reached from any initial patch distribution.

For example, suppose the landscape has been clearcut, so that all the patches are in the open state. Then the initial stage vector is:

$$\mathbf{s}(0) = [500 \ 0 \ 0 \ 0] \qquad \text{Expression 8.8}$$

Figure 8.2 shows the trajectory for both initial vectors. You can see in Figure 8.2a that after approximately 5 time steps, the initial vector has settled into the equilibrium state. If all the patches start in the open state (Figure 8.2b), the same equilibrium is reached. Thus, the equilibrium does not depend on the initial starting conditions.* Instead, it is determined entirely by the transition matrix **A**.

We've seen that the equilibrium vector does not depend on the initial patch vector, but only on the transition matrix itself. Using the rules of matrix algebra, we can solve for the equilibrium vector by calculating the right-hand eigenvector of the transition matrix. Unfortunately, eigenvectors are difficult to calculate by hand for all but the smallest matrices, and the calculation does not provide much insight into the equilibrium numbers.

A much better approach is to program the multiplication directly. This is fairly easy to do on a computer with spreadsheet software (see Donovan and Welden 2001 for detailed instructions). Moreover, this "spreadsheet calculator" can be modified to include additional factors in the model that cannot then be readily solved by calculating eigenvectors. We will use the multiplication approach for all of the examples in this chapter.

STAGE VECTORS AND TRANSITION MATRICES: TWO INTERPRETATIONS

Let us think carefully about how the stage vector and transition matrix are defined and interpreted. In the example we have been developing, the stage vector represents the number of patches in a particular state across a land-

---

*Mathematicians call this property of convergence **ergodicity**. A system is ergodic if its eventual behavior is independent of its initial state (Caswell 2001). All of the simple matrix models of succession exhibit this ergodic behavior. A closely related property is **homogeneity**. A transition matrix is homogenous if its elements do not change through time. It is best to think of homogeneity as a simplifying assumption of the model, and ergodicity as a mathematical consequence of that assumption. Later in this chapter we will talk about some matrices that are not homogenous and consequently may not be ergodic.

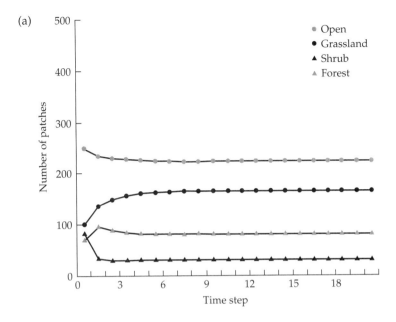

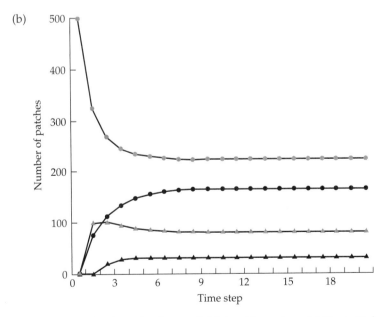

Figure 8.2 Successional trajectories for two initial patch vectors. (a) The initial vector is **s**(0) = [250 100 80 70]. (b) The initial vector is **s**(0) = [500 0 0 0]. The transition matrix **A** for both models is Expression 8.2. Regardless of the starting configuration, all initial vectors converge to a common equilibrium with this transition matrix.

scape. In this case, the elements of the transition matrix describe the fraction of patches that change from one state to the next in each time step.

However, we can also interpret this model on the level of an individual patch. In this case, the transition matrix represents the probability that a patch changes from one state to the next. But what does the stage vector represent? The starting vector would have a single 1 for the patch state that was current; the rest of the vector would be filled with zeroes. At equilibrium, the stage vector represents the fraction of time that a patch spends in each of the different states. Thus, at equilibrium, 16% of the patches (80.75/500) will be in the forest state, and any individual patch will be in the forest state 16% of the time.

Remember that the system is not static just because equilibrium is reached. Individual patches are continuously turning over and changing state, but the frequencies of occurrence of those states is constant. Finally, note that we must assume that the patches are all similar to one another if we are to safely extrapolate this model to a set of patches in a landscape.

Model Assumptions

The simple matrix model of succession—and its ergodic behavior—rests on the following assumptions:

✔ **Communities can be represented as discrete states.** These states should be meaningful classifications of communities into distinct types. The states are mutually exclusive and all-encompassing so that, no matter how the community changes, it can always be assigned to exactly one of the states at any time.

✔ **Time is measured in discrete, evenly spaced units.** The model assumes that time is measured on a discrete scale, and that the time step is relevant and appropriate for the system being modeled.

✔ **The transition matrix is homogenous.** By this we mean that the matrix is constant and does not change from one time step to the next. Although the vector elements do change by the matrix multiplication, the transition matrix itself is invariant.

✔ **No spatial structure.** Transition probabilities do not depend on the spatial arrangement of patches. In other words, the probability of change does not depend on the identity of neighboring patches.

✔ **No density dependence.** Transition probabilities do not change if a patch type becomes very rare or very common.

✔ **Large number of patches.** The model carries over "fractional patches" in its calculations, so it is not affected by "demographic stochasticity" of small patch numbers.

✔ **No time lags.** The changes in the vector are instantaneous and depend only on the current patch state. The changes in the vector do not depend on any previous states, nor do they depend on the path by which the current state was reached.

Model Variations

Some authors have suggested that this list of assumptions is so unrealistic and restrictive that matrix models of succession have not been very useful (Facelli and Pickett 1990). As always, the solution is to carefully modify those features of the model that are thought to be important and see how that additional complexity affects the interpretations and results. But first let us return to the historical ideas on how succession occurs.

SUCCESSIONAL MODELS REVISITED

We have learned how to set up a matrix model to describe how a community changes among a set of patches, and we have seen that a unique equilibrium vector exists that is determined entirely by the transition matrix and not at all by the initial distribution of patch states.

Now we can revisit the three verbal models of succession introduced earlier in this chapter: facilitation, inhibition, and tolerance. What would the transition matrix look like for each of the models? We will use the open–grassland–shrub–forest matrix to suggest appropriate coefficients that describe each of these models.

FACILITATION MODEL

For the facilitation model, we will suppose that the sequence of communities is open → grassland → shrub → forest, with forest representing the climax state. Also, recall the "rules" of the facilitation model: stages cannot be "skipped," and each stage must occur in sequence to facilitate the change to the next state. Our transition matrix for this model might look like this:

		Stage at time t			
		Open	**Grassland**	**Shrub**	**Forest**
Stage at time t + 1	**Open**	0.10	0.10	0.10	0.01
	Grassland	0.90	0.10	0.00	0.00
	Shrub	0.00	0.80	0.10	0.00
	Forest	0.00	0.00	0.80	0.99

Expression 8.9

In this example, we have assumed that, for any patch, there is always a 10% chance that a disturbance will re-set the system. However, the "climax" forest community is disturbed relatively infrequently, and there is only a 1% chance of moving from the forest to the open state in each time step. In the grassland and shrub patches, we have assumed that there is only a 10% chance of remaining in the same state, and an 80% chance of "graduating" to the next stage in the facilitation sequence. The remaining 10% (100 – 80 – 10) represents the chance of a disturbance re-setting the system.

This facilitation model is defined by the unique arrangement of zeroes in the matrix, which ensures that patches move through the sequence in orderly fashion. The corresponding loop diagram is shown in Figure 8.3a.

INHIBITION MODEL

For the inhibition model, we will assume that each of the three community states (grassland, shrubland, forest) can be replaced by another community state only through an intervening disturbance that frees up space. The transition matrix might look like this:

		Stage at time t			
		Open	**Grassland**	**Shrub**	**Forest**
Stage at time t + 1	**Open**	0.10	0.10	0.10	0.10
	Grassland	0.30	0.90	0.00	0.00
	Shrub	0.30	0.00	0.90	0.00
	Forest	0.30	0.00	0.00	0.90

Expression 8.10

Figure 8.3 Idealized loop diagrams for simple succession models. (a) Facilitation model. (b) Inhibition model. (c) Tolerance model.   ▶

(a)

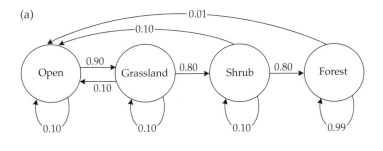

(b)

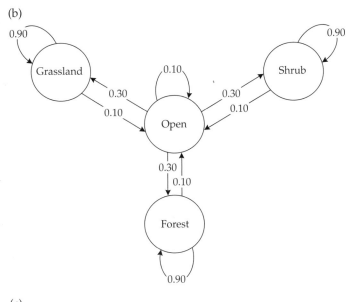

(c)

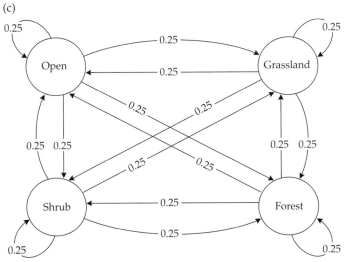

In this matrix, there is a 10% chance of a disturbance in each time step, regardless of the initial state. Once a community is established, it stays intact unless it is removed by a disturbance. Following a disturbance, there is an equal chance (30%) of entering any of the community states, and only a 10% chance of remaining in the open state for two time steps in a row. The corresponding loop diagram is shown in Figure 8.3b.

TOLERANCE MODEL

For the tolerance model, we will assume that all states—including the open state—are equally likely. This generates a transition matrix with identical transition elements:

		Stage at time t			
		Open	**Grassland**	**Shrub**	**Forest**
	Open	0.25	0.25	0.25	0.25
Stage at time t + 1	**Grassland**	0.25	0.25	0.25	0.25
	Shrub	0.25	0.25	0.25	0.25
	Forest	0.25	0.25	0.25	0.25

Expression 8.11

In this model, each community neither inhibits nor facilitates replacement by other communities. Therefore, all the transitions are equally likely. The corresponding loop diagram is shown in Figure 8.3c.

MODEL COMPARISONS

To compare the predictions of these different transition matrices, we began with a system in which 1000 patches were initially in the open state. The trajectories of the three systems are rather different. The inhibition and tolerance matrices settle into their equilibria after a single time step, whereas the facilitation matrix reaches equilibrium after approximately 20 time steps (Figure 8.4).

As we pointed out, these models are all ergodic, and quickly settle into their characteristic equilibria. If we know the elements of the transition matrix, we can easily forecast that equilibrium. But the converse is not true. If we know the equilibrium distribution, we cannot infer a unique transition matrix, because there are many different matrices that can lead to the same pattern. Moreover, if we were trying to construct the transition matrix using only the distribution of patch states in nature, we would have to further assume that the system had already reached its equilibrium state.

It seems that the best way to distinguish among the facilitation, inhibition, and tolerance models is by directly comparing the structure of the transition

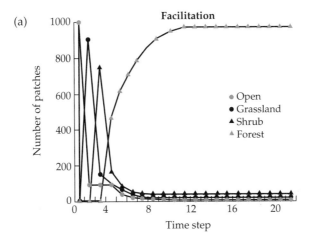

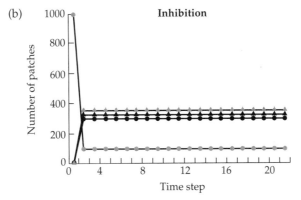

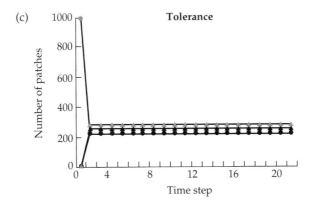

Figure 8.4 Patch trajectories for simple successional models. In each model the initial vector is **s**(0) = [1000 0 0 0]. (a) Facilitation model. (b) Inhibition model. (c) Tolerance model.

matrices themselves. In other words, the assumptions of the models are perhaps more distinct than the predictions (Connell et al. 1987). Nevertheless, constructing a transition matrix and using it to forecast the stage distribution of a community does provide insight into how successional change occurs, as we will see in McAuliffe's (1988) study of change in desert vegetation.

OTHER MODELS

Many interesting variations can be created from our basic matrix model by altering its assumptions and introducing more complexity. Here we will only sketch out the possibilities without elaborating the details. If the successional process has a "memory," the transitions may depend not only on the current state, but also on the state of the community several time steps in the past. As in population growth models with time lags (see Chapter 2), this matrix model can exhibit complex transient dynamics, as we will see in the study of coral reef dynamics by Tanner et al. (1996).

Another possibility is that the transition matrix itself is not constant, but changes at each time step. For example, each matrix element might be sampled from a distribution with a mean and variance, as in our stochastic models of exponential growth (see Chapter 1). In this case, there is no longer a simple equilibrium state, although the assemblage will usually converge on a sort of "average" or expected distribution with time (Caswell 2001).

Alternatively, the matrix elements may change through time in a systematic way, reflecting long-term environmental changes such as global warming or increasing nitrogen deposition. In these models, the assemblage may never settle into a steady state, and the appearance of the landscape will change substantially as different kinds of sequential matrices operate on the stage vector (Doak and Morris 1999).

Making the successional model spatially explicit forces us to use a completely different kind of mathematics to describe community change. It is no longer sufficient to multiply a transition matrix by a stage vector. Instead, we must model the individual patches themselves and keep track of their individual states. In this sort of model, the rules for state transitions may depend on the current state of adjacent cells. For example, a simple rule might be that the probability of a transition into a particular state is proportional to the number of adjacent cells in that state (Molofsky 1994). Such a rule would describe the realistic scenario in which most of the propagules that enter a patch come from nearby sites. These so-called **cellular automata models** can generate very complex dynamics,* resulting in local extinction, patchiness, and spatial "waves" of pattern that pass through a landscape (Wolfram 1984; Durrett and

---

*The dynamics may resemble those seen in the ancient Japanese board game "Go," in which a player captures an opponent's pieces by surrounding them with his own.

Levin 1994). Finally, computationally intensive **individual-based models** keep track of the birth, growth, and death of individuals and may incorporate mechanisms of dispersal, facilitation, growth, and inhibition in a spatially explicit landscape. These highly realistic models have been fit to large data sets on succession in temperate forests (Botkin 1992, Pacala et al. 1996).

All of these models generate complex and interesting patterns. However, it is difficult to collect field data that can be used to fit the model parameters, so some of these exercises are only of theoretical interest. As we will see in the following two examples, it takes a good deal of hard work and creativity just to estimate the parameters for the simplest matrix model of succession.

Empirical Examples

MARKOVIAN DYNAMICS OF DESERT VEGETATION

If you ever "take that California trip" and drive across the Mojave Desert, you will pass through thousands of hectares of open space dominated by creosote (*Larrea*) and wormwood (*Ambrosia*). Although the desert landscape seems static, it is actually a dynamic system that undergoes successional change, though very slowly. McAuliffe (1988) developed a simple three-state Markov model to describe transitions between *Larrea*, *Ambrosia*, and open space.

Because desert plants decay very slowly after they die, it is possible to estimate mortality and patch transitions with careful field censuses, and to measure growth rings in the stems of the shrubs. McAuliffe (1988) combined these data to construct the transition matrix for desert communities around San Luis, Arizona (Yuma County) in the Mojave Desert. Figure 8.5 shows the transition matrix that he estimated for the San Luis site. The matrix describes a vector with three states (open, *Larrea*, *Ambrosia*) and has a time step of 1 year.

As you can see, the matrix does not clearly resemble any of the idealized matrices we proposed for the facilitation, tolerance, and inhibition models. However, *Larrea* rarely colonizes open space, and seedlings of *Larrea* are almost always found beneath the canopy of *Ambrosia*. This is a type of facili-

		Stage at time t		
		Open	Ambrosia	Larrea
Stage at time t + 1	Open	0.99854	0.031	0.0016
	Ambrosia	0.0013	0.96842	0
	Larrea	0.00016	0.00058	0.9984

Figure 8.5 Transition matrix for desert plant communities at San Luis site in the Mojave desert. (Data from McAuliffe 1988.)

tation, although it does not lead to orderly species replacement as in the classic facilitation model. Also, because of the slow rates of change in this system, *Larrea* does not immediately replace *Ambrosia*, so that transition is set to 0.0 for the time step of one year in this model. The diagonal elements of this matrix all have values very close to 1.0. As we will see, this will have an important effect on the dynamics.

In addition to estimating the transition matrix, McAuliffe (1988) also measured the proportion of the landscape that was occupied by each of the three states. Next, he used the transition matrix to forecast the equilibrium frequencies of each state. Note that these are two distinct kinds of data: indirect measurements of state changes and species persistence for constructing the transition matrix, and direct measurements of patch occupancy for estimating the stage vector.

How well does the observed stage vector match up with the equilibrium stage vector that is predicted by the measured transition matrix? The observed and expected frequencies match fairly well (Figure 8.6). However, the model predicts that *Larrea* should occupy 9.9% of the patches, whereas only 2.8% were observed. McAuliffe (1988) suggested that density-dependent mortality might be important in reducing *Larrea* cover below the model predictions.

The forecasting model also provides some insight into how desert communities can be expected to recover from human disturbances. In the desert, small-scale human disturbances include the activities of cactus collectors and ATV enthusiasts, whereas large-scale disturbances include the construction of giant military bases and the relentless sprawl of new housing developments. We can use McAuliffe's (1988) model to ask how long it will take for

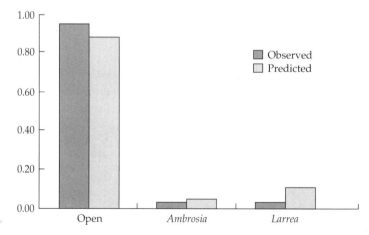

Figure 8.6 Observed and expected frequencies of patch states for desert plant communities at San Luis. (Data from McAuliffe 1988.)

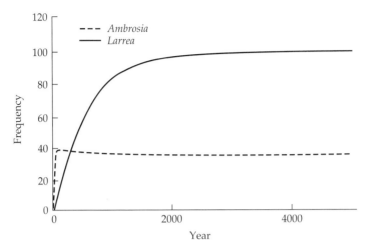

Figure 8.7 Simulated increase in *Larrea* and *Ambrosia* patches. The transition matrix is given in Figure 8.5. The initial stage vector consisted of 1000 open patches.

Larrea and *Ambrosia* cover to reach equilibrium after a 100% clearing. Figure 8.7 illustrates the trajectories for the two species, beginning with 1000 patches that are in the open state.

The answer is that it takes an extremely long time—over 2000 years—for this system to settle into its equilibrium. The approach to equilibrium is slow because the diagonal values of the transition matrix are all very close to 1.0. In other words, very few patches actually change states from one year to the next. Consequently, recovery from disturbance (whether natural or human-caused) is very slow.

MODELS OF CORAL REEF SUCCESSION

Ecologists need long-term data to estimate transition probabilities and construct realistic models of succession. Since 1962, the ecologist Joseph Connell and his colleagues have conducted regular censuses of coral communities in a set of three permanent 1-m^2 quadrats at Heron Island in the Great Barrier Reef of Australia. Each census consists of a photograph and measurement of the transitions that occur beneath a 20×20 square grid superimposed on the photographs. Quadrats were censused approximately once every 19 months, for a total of 19,200 transitions.

Tanner et al. (1996) used these data to develop realistic matrix succession models. First, they grouped the 72 coral and 9 algal species into 6 categories of hard corals, 1 category of soft coral, and 1 category of algae. They also included the state of open space, for a total of 9 states in their matrix. Next, they developed four different kinds of matrix models:

Model 1: First-order Markov model. This model is identical to the basic model we developed in this chapter. A single 9×9 transition matrix was estimated from all of the data, and that transition matrix was applied to the 9-element stage vector each year. Because there is no obvious competitive dominant in this community, the transition matrix contained relatively few zeroes (see Problem 8.1) and is qualitatively similar to a tolerance model.

Model 2: Second-order Markov model. This model incorporates the idea that transition probabilities might depend not only on the current state, but also on the state one time step previous time. For example, the transition from alga to free space might depend on whether the patch history was (alga, alga), or (soft coral, alga). In other words, the probability depends not just on the current state (alga), but also on the state in the previous time step (soft coral or alga). Thus, the system has a "memory" of one time step, so that the transition probabilities are determined by the two-step sequence of occupancy.

Models 3, 4: Semi-Markov models. These are similar to the second-order model. However, transition probabilities depend on the absolute amount of time that a patch has been occupied. Each species has a characteristic "waiting time," and once that amount of time has passed, there is a transition to a new state. In Model 3, a complete distribution of waiting times of between 1 and 11 steps was estimated from the data. In Model 4, only 2 time steps were used to estimate waiting times. These models take into account both the different life histories of the component species and the fact that ephemeral species will not hold patches beyond a certain point, regardless of the estimated transition probabilities in the simple model.

How do the predictions of these models differ? For soft corals, the two semi-Markov models predicted higher occupancy than the first- or second-order Markov models. The more complex models exhibited a longer period of transient dynamics before settling into their equilibrium states. However, the main result is that the overall predictions of the complex models were surprisingly similar to the predictions of the simple first-order Markov model (Figure 8.8). Tanner et al. (1996) suggested two reasons for the good match of the simple and complex models. First, a rapid turnover of coral colonies and algae means that only a small proportion of colonies survive long enough for historical effects to be important. Second, these coral communities are disturbed by tropical cyclones far more frequently than the estimated time to equilibrium, ensuring that historical effects are minimal.

Other long-term studies of this sort are needed, but these results suggest that the basic Markov model might be a useful forecasting tool, even if it does not incorporate subtleties such as historical effects and individual life histories.

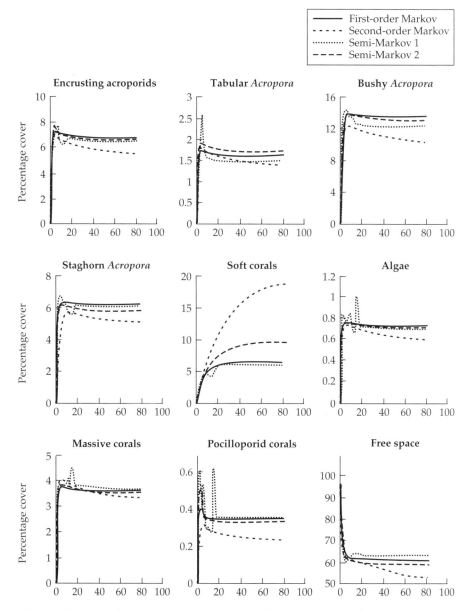

Figure 8.8 Successional trajectories for coral reef communities in the Great Barrier Reef, Australia. See text for details of the four Markov models. (Adapted from Tanner et al. 1996.)

Problems

8.1 Here is the transition matrix that Tanner et al. (1996) measured for their simplest (first-order) Markov model:

	Stage at time t								
Stage at time t + 1	***Pocilloporid corals***	***Massive corals***	***Algae***	***Soft corals***	***Staghorn Acropora***	***Bushy Acropora***	***Tabular Acropora***	***Encrusting acroporid corals***	***Free space***
Encrusting acroporid corals	0.354	0.046	0.032	0.032	0.000	0.071	0.025	0.039	0.059
Tabular Acropora	0.021	0.314	0.005	0.004	0.003	0.014	0.000	0.000	0.014
Bushy Acropora	0.066	0.030	0.478	0.082	0.016	0.090	0.076	0.105	0.091
Staghorn Acropora	0.049	0.016	0.038	0.439	0.009	0.057	0.031	0.053	0.039
Soft corals	0.001	0.005	0.005	0.004	0.835	0.005	0.011	0.000	0.014
Algae	0.009	0.036	0.007	0.004	0.000	0.033	0.015	0.000	0.007
Massive corals	0.015	0.003	0.013	0.014	0.006	0.052	0.340	0.000	0.032
Pocilloporid corals	0.002	0.005	0.001	0.001	0.000	0.000	0.000	0.224	0.004
Free space	0.482	0.544	0.421	0.421	0.131	0.678	0.501	0.579	0.741

a. What is the probability of a transition from tabular *Acropora* to bushy *Acropora*?

b. Which stage is most likely to be replaced by free space once it has colonized?

c. Which transitions were never observed in this community?

*8.2 Suppose there are 900 patches, and the community begins with exactly 100 patches in each of the 9 states. How many patches will be in each state at the next time step?

*Advanced problem

Appendix

Although many ecology courses require calculus as a prerequisite, my experience has been that students have a hard time relating the material they learned in their calculus class to ecological models of population growth. This appendix explains the process of building ecological models and how calculus is used to do this.

CONSTRUCTING A POPULATION MODEL

First things first. What, exactly, are we trying to do when we build a population model? Our goal is to write an equation, or function, that will tell us what the population size N will be at some future time t:

$$N = f(t) \qquad \text{Expression A.1}$$

If we plug into this function f the elapsed time t, the function will predict the population size N at time t. We will obviously need more information than just t in order to build this function. The size of the population will depend on many things other than the amount of time that has elapsed. Much of the text of this primer is devoted to elaborating the biological and physical details of the model.

THE DERIVATIVE: THE VELOCITY OF A POPULATION

Although our goal is to produce a function like Expression A.1, this is not the place we actually begin building a population model. The reason for this is that it is difficult to directly model factors that cause populations to be large or small. Instead, it is much easier to model factors that cause populations to *increase* or *decrease* in size.

Thus, we make an important distinction between the size of a population (N), and its current growth rate (dN/dt). The size of the population is the total number of individuals, and it is measured in units of individuals. The growth rate is the "velocity" of the population, that is, the change in population size per unit time, and it is measured in units of individuals/time.

How, exactly, do we measure the growth rate of a population? Suppose we census a population of parrots and count 500 birds. We return to our population one year later and count 600 birds. The population growth rate can be measured as the change in population size divided by the change in time:

$$\text{Population growth rate} = \frac{(600 - 500)}{(1 - 0)} = 100 \text{ birds/year} \quad \text{Expression A.2}$$

This measurement is an average and is valid as long as the population is growing at a constant rate during the year. However, it is likely that the rate of population growth changes with time. In other words, if we measured the population growth rate during the breeding season, we would measure a rate much greater than 100 birds/year. And, if we measured population growth at a time of the year when females were not giving birth, our measured rate of population growth would be zero, or even negative if there had been losses due to death or emigration.

In a similar way, if you took a car trip, you could calculate your average velocity by dividing the distance travelled by the amount of time elapsed. However, during the course of your journey, the speed of your car may have been very different from this average as you drove through traffic or on open stretches of highway.

Thus, velocity is best measured over a very short time interval. Suppose we measure the population size at time t, and then again some short time later at $t + x$. Our equation for population growth rate becomes:

$$\text{Population growth rate} = \frac{(N_{t+x} - N_t)}{(t + x - t)} \quad \text{Expression A.3}$$

The derivative of any function is simply this velocity measured over an infinitely small interval. In other words, x is very small, so small that it approaches zero. Writing this as a continuous differential equation, we have:

$$\text{Population growth rate} = dN / dt \quad \text{Expression A.4}$$

Here, dN means the change in population size measured over a very short interval dt. Graphically, dN/dt turns out to be the slope of our original function $f(t)$. Thus, if we plot population size N, as a function of time t, the population growth rate at any time t is the line tangent to the function at that point t (Figure A.1). Notice that the slope of this line may change depending on where on the population graph we measure it. In other words, the population growth rate is not a constant, but may change with time.

It is critical that you grasp the distinction between the population size (N) and the population growth rate (dN/dt). Population size will always be a non-negative number, but population growth rate could be positive, negative, or zero, depending on whether the population is increasing, decreasing, or unchanging. In fact, it is often the case that large populations will have low growth rates, and vice versa. Chapter 2 develops this model of density-dependent population growth in detail.

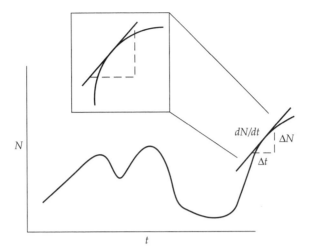

Figure A.1 Population size (N) plotted as a function of time (t). At any value of t, the slope of the line that is tangent to the curve at that point is the population growth rate, dN/dt.

MODELING POPULATION GROWTH

Having defined the population growth rate as the variable of interest, the next step is to begin incorporating the details of population growth into the model. As you look through this primer, you will see that the population growth rate is defined as a series of terms, some of which are positive and increase population growth, and some of which are negative and decrease growth.

Each term in the model carries with it certain assumptions about how populations are growing. The way to build new models is to relax or alter those assumptions, and then modify the equation to reflect the new scenario. For example, Equation 1.1 is the differential equation for exponential population growth:

$$dN/dt = rN \qquad \text{Equation 1.1}$$

As explained in Chapter 1, one of the assumptions of Equation 1.1 is that there is no migration in or out of the population. We can now relax this assumption and modify Equation 1.1 to incorporate migration.

When we model population growth, we must be careful in deciding whether the growth factors are behaving as constants, or as factors that are proportional to population size. For example, suppose that a constant number of parrots emigrated from our population each year (c). This rate c does not change with time or population size. Its units are number of emigrants/year.

Thus, our new model of population growth is:

$$dN/dt = rN - c \qquad \text{Equation A.1}$$

There are other ways we could describe the emigration process. For example, suppose a certain fraction of the population departs as emigrants at a constant rate. Let the constant g represent the per capita or per individual rate of emigration. The units of g are individuals/individual • time. Therefore, the emigration rate (emigrants/time) is gN. This model of population growth is:

$$dN/dt = rN - gN \qquad \text{Equation A.2}$$

As we will see, these two models make rather different predictions about population growth.

SOLVING FOR THE EQUILIBRIUM

Now that we have our population growth equation defined, the next step is to solve for the equilibrium. We are interested in determining when, if ever, our model population stops growing. In other words, will the population reach a size beyond which it neither increases nor decreases? Mathematically, this equilibrium corresponds to the population size for which $dN/dt = 0$. Note that there may be more than one equilibrium point, that is, more than one population size that satisfies this condition. Returning to Equation A.1, we can set it equal to zero and solve for equilibrium:

$$0 = rN - c \qquad \text{Expression A.5}$$

$$c = rN \qquad \text{Expression A.6}$$

$$N = c/r \qquad \text{Equation A.3}$$

Equation A.3 gives the equilibrium for Equation A.1, in which a population that is growing exponentially is also losing emigrants at a constant rate. Equation A.3 says that our model population will stop growing if the population size equals the ratio c/r.

The equilibrium for Equation A.2 is rather different. In this model, a population that is growing exponentially loses a constant proportion of individuals as emigrants. Setting this equation equal to zero and solving gives us:

$$0 = rN - gN \qquad \text{Expression A.7}$$

$$gN = rN \qquad \text{Expression A.8}$$

$$g = r \qquad \text{Equation A.4}$$

Notice that, in this solution, there is no unique value of N. In other words, there is no particular population size for which growth will cease. Instead, the condition for zero growth is that the rate of migration (g) is exactly balanced by the instantaneous rate of increase (r). No matter how large or how small the population, growth will cease if these two constants equal each other. Recalling that the instantaneous rate of increase (r) is the difference between the instantaneous birth (b) and death (d) rates (see Chapter 1), we have:

$$b = g + d \qquad \text{Expression A.9}$$

In other words, growth will cease if the birth rate equals the sum of the emigration rate and the death rate. If b is greater than $g + d$, the population will increase exponentially, and if b is less than $g + d$, the population will decrease exponentially.

ANALYZING THE STABILITY OF THE EQUILIBRIUM

Although we have now defined the equilibrium solutions for our models, there is one additional step that is important. We have to analyze their stability. As described in the text, there are three sorts of stability that a point equilibrium may acheive: stable, unstable, or neutrally stable.

To understand the idea of stability, imagine a population is at equilibrium, so that the positive and negative forces affecting population growth are in balance. Now we perturb our population by either adding or removing a few individuals. What happens now that the population no longer is in equilibrium? If it returns to the original population size, the equilibrium point is stable. If the population continues to move away from its initial size, the equilibrium point is unstable. Finally, if the population remains at rest at the new population size, the equilibrium point is neutrally stable.

Technically, this analysis is based on the idea of local stability, because we are only analyzing perturbations that take the population a short distance away from the equilibrium point. If the population is perturbed more violently, it won't necessarily return to a locally stable equilibrium. If the perturbation pushes the population close to a different equilibrium point, which is also locally stable, the population may move towards that new equilibrium.

How can we test mathematically for the stability of the equilibrium? The simplest approach is to first solve for the equilibrium point as we did above. Next, we set the population size slightly above this equilibrium, and use this value of N to solve for the growth equation. Is the population growth rate positive when the population is slightly above equilibrium? If so, it will continue to move away from the equilibrium point, which is therefore unstable. On the other hand, if the population growth rate is negative when the population is slightly above equilibrium, the equilibrium point is probably stable. Finally, if the population growth rate is also zero when the population is displaced slightly above equilibrium, we have a neutral equilibrium. We should also substitute a value for N that is slightly below the equilibrium population size to make sure the behavior is consistent. In this case, the population should rise towards a stable equilibrium, continue to decrease away from an unstable equilibrium, and remain in place ($dN/dt = 0$) for a neutral equilibrium.

For simple growth equations, we can also analyze stability graphically. To do this, we take the positive and negative growth elements of the equation, and divide them each by N, so they are expressed as per capita (per individual) rates. Next, we plot these two curves on a graph with population size (N) on the x axis and the per capita growth elements on the y axis. If the two curves do not cross, there is no equilibrium because there is no population size for which the per capita birth and death rates are equal. If the curves intersect in a portion of the graph where the population size is greater than or equal to zero, the intersection constitutes an equilibrium solution to the equation. The point on the x axis of the graph directly beneath the intersection represents the equilibrium population size. If there is more than one intersection, then there is more than one equilibrium point. And, if the two curves are identical to one another, then all population sizes are equilibrium points.

You should confirm that the intersection of the points represents an equilibrium by setting the growth equation equal to zero and solving it algebraically. Next, we examine the birth and death curves for populations slightly displaced from the equilibrium. If the birth curve lies above the death curve, then the population will increase at that population size. If the birth curve lies below the death curve, the population will decrease. We can draw these increases or decreases as horizontal arrows on the graph with the origin of each arrow placed just to the left or to the right of each equilibrium point. These graphs will quickly show us whether the equilibrium point is locally stable or unstable. Remember that if the population neither increases nor decreases when displaced from equilibrium, the equilibrium point is neutrally stable. This will be the case when the birth and death curves are equal to one another, because a slight movement to the left or right of the equilibrium will bring the population to another equilibrium.

An example of this analysis is shown in Figure 2.1, which illustrates the per capita birth and death functions plotted against population size for the logistic growth model. In the logistic model, the per capita birth rate decreases and the per capita death rate increases as population size increases. These two curves intersect at an equilibrium point.

As Chapter 2 explains, this equilibrium point corresponds to the carrying capacity, K, of the environment. K is a stable equilibrium point. At population sizes smaller than K, the per capita birth rate exceeds the per capita death rate, so the population increases. Above K, the death rate exceeds the birth rate, so the population declines towards K.

Now let us analyze Equation A.2, in which there is a constant per capita rate of emigration out of the population ($dN/dt = rN - gN$). In this case, the equilibrium solution is $r = g$ (Equation A.4). To analyze the stability of the equilibrium, we plot as a function of population size the per capita elements of the equation that lead to population increases and decreases. The growth term in this equation is rN, but plotting it on a per capita basis gives $rN/N = r$. Similarly, the emigration term is plotted as g on a per capita basis. If r and g do not equal each other, the plot is of two parallel lines that never cross (Figure A.2). If r is greater than g, the population is increasing exponentially,

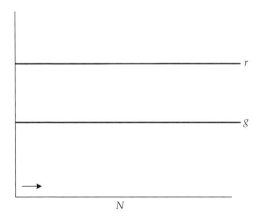

Figure A.2 The intrinsic rate of increase (r) and the proportional rate of emigration (g) plotted as a function of population size (N). Because these two curves never cross, there is no population size for which these two forces are in balance, and hence no equilibrium population size for which $dN/dt = 0$.

and if r is less than g, the population is decreasing exponentially. In both cases, an equilibrium is never reached.

However, if r equals g, the two lines are identical and represent a neutral equilibrium. In this case, any population size that is chosen will meet the condition of $r = g$. If the population is displaced above or below its equilibrium, the new population also corresponds to a point at which r equals g, and there will be no further population growth. Thus, we have a neutral equilibrium.

For a more complex example, we will now analyze the equilibrium of Equation A.1. In this model, there is exponential growth (rN) and emigrants are leaving the population at a constant rate c. The equilibrium solution is a population size of $N = c/r$ (Equation A.3).

To analyze the stability of this equilibrium, first plot the two components of growth on a per capita basis as a function of N. As before, plotting the growth term rN on a per capita basis means plotting $rN/N = r$ on the graph.

You might be tempted to plot the emigration rate c as a straight line as well, but this would be incorrect. Remember that each term must be plotted on a per capita, or per individual, basis. This means that you must plot c/N as a function of N, which is the graph of a hyperbola.

There are two cases to consider. The first is when the instantaneous rate of increase (r) is less than zero. In this case, the two curves do not intersect, and there is no equilibrium.[*] This makes intuitive sense. If the population is decreasing exponentially, the additional loss of emigrants at a constant rate will not lead to an equilibrium, and the population will continue to decline towards zero.

However, if the intrinsic rate of increase is positive, the straight line graph of r lies above the x axis and intersects the hyperbolic curve of c/N at the value c/r, which establishes the equilibrium (Figure A.3). Is this equilibrium point stable? To the left of the intersection, the curve of c/N is above the r line, so the population will decrease. To the right of the intersection, the positions are reversed, so the population will increase. In other words, any movement away from the equilibrium point takes the population even further from the intersection. Thus, the population has an unstable equilibrium point. If the population declines, even slightly below this equilibrium, emigration will overpower growth, and the population will decline towards zero. If the population is displaced above its equilibrium, it will begin increasing.

---

[*]Technically, there is an equilibrium point at $N = 0$, which is true for most population growth models. The zero equilibrium point may be stable or unstable, depending on whether a very small population will grow or shrink. As the Appendix explains, the zero equilibrium point for Equation A.1 is stable. Some population models even generate equilibria with negative population sizes. These equilibria are of interest to mathematicians, but not to ecologists.

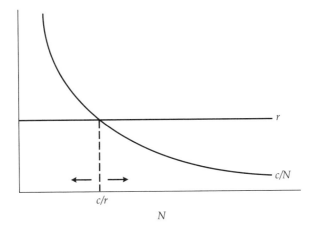

Figure A.3 The intrinsic rate of increase (r) and the constant rate of emigration (c) plotted as a function of population size (N). Because the curves are plotted on a per capita basis, the emigration rate is plotted as c/N. The two curves intersect at the equilibrium point $N = c/r$. Because the population growth rate is positive to the right of the intersection and negative to the left of the intersection, the equilibrium is unstable.

Thus, introducing a constant rate of emigration from a population that is increasing exponentially (positive r) results in an unstable equilibrium. This unstable equilibrium represents a minimum viable population size, because if the population dips below the equilibrium, it will continue decreasing towards zero.

These analyses illustrate graphical and numerical methods that may be used to solve for the equilibrium in simple models of population growth and to analyze the stability of the equilibrium. These analyses also demonstrate why it is important to state the conditions of the model carefully. Whether we modeled emigration as a constant rate (c) or as a constant per capita rate (g) had a big influence on the predictions of the model. For an example of a population growth model that has more than one equilibrium point, one of which is stable and one of which is unstable, see Problem 2.3.

For models that incorporate more than one interacting species (Chapters 5 and 6), the approaches used here will not work, because the equilibrium population size for one species will depend on the population size of the other species, and vice versa. In these cases, a state-space graph can often be used to analyze equilibrium states. In a state-space graph, species' isoclines are plotted on a graph in which the abundances of the two species form the x and y axes. The method is described in detail in Chapter 5. These graphical solutions are appropriate for simple models with linear isoclines. However, for non-linear models (e.g., some of the model variations in Chapter 6) it is not always possible to infer the dynamics from the state-space graph. More complex mathematical tools are needed, and Roughgarden (1998) provides a good introduction to the methods. Also, some computer mathematics programs, such as *Mathematica* and *Matlab*, will solve equations numerically and analyze their stability.

THE INTEGRAL: PROJECTING POPULATION GROWTH

Let us review what we have done up to this point. First, we modeled population growth by describing mathematically the components that determine the population growth rate (dN/dt). Next, we set this growth equation equal to zero and solved for the equilibrium point(s). Finally, we used simple graphical and numerical analyses to determine whether the equilibrium was unstable, stable, or neutral.

The chapters in the primer follow a similar strategy for the analysis of several kinds of population models. Although the text presents differential equations for population growth and the algebraic manipulations necessary to solve these equations, no calculus is actually used!

So, is there a need for calculus at all in modeling population growth? Recall our original goal stated at the start of this appendix: to derive a function that would allow us to predict population size at a given time t. What we have done instead is to derive an equation that will give us the population growth rate (dN/dt) at a given time or given population size.

Because integration is the inverse operation of differentiation, the integral of a growth equation will forecast population size. Hence, the rules of integration can be applied to growth equations to convert them to a form that can be used to predict or forecast population size. As an example, here are the steps of integration for the exponential growth equation $dN/dt = rN$:

$$dN_t / dt = rN_t \qquad \text{Equation 1.1}$$

$$dN_t = rN_t dt \qquad \text{Expression A.10}$$

$$dN_t / N_t = rdt \qquad \text{Expression A.11}$$

Recall from your calculus that the derivative $d\ln(x)/dx = 1/x$, so that $dx/x = d\ln(x)$. Thus:

$$d\ln(N_t) = rdt \qquad \text{Expression A.12}$$

Integrating both sides gives:

$$\ln N_t - \ln N_0 = rt - rt_0 \qquad \text{Expression A.13}$$

Rearranging the terms gives:

$$\ln(N_t / N_0) = r(t - t_0) \qquad \text{Expression A.14}$$

$$N_t / N_0 = e^{r(t-t_0)} \qquad \text{Expression A.15}$$

$$N_t = N_0 e^{r(t-t_0)} \qquad \text{Expression A.16}$$

Substituting $t_0 = 0$ gives:

$$N_t = N_0 e^{rt} \qquad \text{Equation 1.2}$$

Mathematically, both equations say the same thing, but one is a function that gives the population growth rate as its output (Equation 1.1), and the other is a function that gives population size as its output (Equation 1.2).

Another place where the rules of calculus come in handy is in the conversion from discrete to continuous equations. For example, Equation 1.5 gives the relationship between λ, the finite rate of increase, and r, the instantaneous rate of increase:

$$e^r = \lambda \qquad \text{Equation 1.5}$$

This derivation is based on the fact that, as n approaches infinity:

$$\left(1 + \frac{x}{n}\right)^{\frac{n}{x}} \to e \qquad \text{Expression A.17}$$

The base e is a constant ($e \approx 2.718$) named in honor of the Swiss mathematician Leonhard Euler (1707–1783), who also gave ecology the Euler equation (Equation 3.13). As explained in Chapter 1, the discrete growth factor r_d is equivalent to the instantaneous rate of increase r if the time step is infinitely small. Also, remember that the finite rate of increase $\lambda = (1 + r_d)$. Letting $r = x/n$, we have:

$$e = (1 + r)^{\frac{1}{r}} \qquad \text{Expression A.18}$$

$$e = (\lambda)^{\frac{1}{r}}$$ Expression A.19

Raising both sides to a power of r gives:

$$e^r = \lambda$$ Equation 1.5

Thus, a thorough knowledge of calculus will make it easier to grasp models of population growth, but it is not essential for understanding the operation of the basic models described in this primer. As we try to make our ecological models more realistic, it becomes very easy to write equations that are so complex that they cannot be solved analytically and must be evaluated numerically. Although the basic models may make unrealistic assumptions, their predictions are simple and can be tested empirically. The equations analyzed in this primer are the foundation of modern population and community ecology, and you should have a solid understanding of how they operate.

Solutions to Problems

1.1. Rearranging Equation 1.3:

$$r = \ln(2)/t = \ln(2)/50 \text{ years}$$
$$= 0.01386 \text{ individuals } / \text{ (individual } \cdot \text{ year)}$$

$N_0 = 5.4$ billion and $t = 7$ years. From Equation 1.2 we have:

$$N_7 = 5.4 \, (e^{(0.01386)(7)})$$
$$= 5.95 \text{ billion humans}$$

1.2. Because there were 400 births, 150 deaths, and no migration, from Expression 1.1:

$$N_{t+1} = 3000 + 400 - 150 = 3250$$

We can arrange Expression 1.15 to give:

$$\lambda = N_{t+1}/N_t$$

Thus, $\lambda = 3250/3000 = 1.0833$

Using Equation 1.6 to convert λ to r gives:

$$r = \ln(1.0833) = 0.0800 \text{ individuals/(individual } \cdot \text{ month)}$$

From Equation 1.2, the population size after six months will be:

$$N_6 = 3000(e^{(0.0800)(6)}) = 4848 \text{ beetles}$$

1.3. First, take the natural logarithms (base e) of the five consecutive population sizes, yielding: 4.605, 5.063, 5.753, 5.986, and 6.677. Next, plot these values as a function of time:

(a)

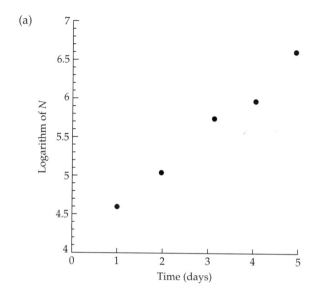

Because resources aren't limited, we can draw a straight line to fit all five of the data points. Although the points don't fall precisely on the line, this line gives a good estimate of population growth:

(b)

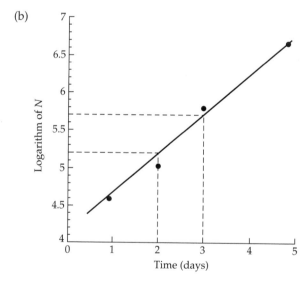

Finally, we measure the slope of this straight line to estimate r. The slope of a line is simply $(\Delta y / \Delta x)$. Using the dashed lines to calculate the slope, we have $(5.7 - 5.2)/(3 - 2) = 0.5$. So, our estimate of r for this flatworm population is:

$r = 0.5$ individuals/(individual • day)

1.4. Because this is an annual species, we need to use the model of discrete population growth with a time step of one year. If the population is increasing by 12% each year, $\lambda = (1 + 0.12) = 1.12$. From Equation 1.5:

$r = \ln(\lambda) = \ln(1.12) = 0.113$ individuals / (individual • year)

Finally, we use Equation 1.3 to calculate the approximate doubling time:

$$t_{double} = \ln(2)/r = \ln(2) / 0.113$$
$$= 6.1 \text{ years}$$

The answer is only approximate because Equation 1.3 is for a continuously growing population.

1.5. For small populations growing with demographic stochasticity, the probability of extinction can be calculated from Equation 1.15. In the undisturbed state, the probability of extinction with a population size of 50 plants is:

$$P_{(extinction)} = \left(\frac{d}{b}\right)^{N_0} = \left(\frac{0.0020}{0.0021}\right)^{50} = 0.087$$

If the shopping mall is built, we have:

$$P_{(extinction)} = \left(\frac{d}{b}\right)^{N_0} = \left(\frac{0.0020}{0.0021}\right)^{30} = 0.231$$

So the proposed development threatens to increase the risk of extinction from about 9% to 23%.

CHAPTER 2

2.1. To solve this problem, we first need to determine N, the population size. From Figure 2.3a, we know that the maximum possible growth rate for a population growing according to the logistic model occurs when $N = K/2$, so $N = 250$ butterflies. Plugging these values into Equation 2.1 gives:

$$\frac{dN}{dt} = rN\left(1 - \frac{N}{K}\right)$$
$$= 0.1(250)[1 - (250/500)]$$
$$= 12.5 \text{ individuals/month}$$

2.2. For a population growing according to the logistic equation, we know that the maximum population growth rate occurs at $K/2$, so K must be 1000 fish in this case. If the population is stocked with an additional 600 fish, the total size will be 1100. From Equation 2.1, the initial instantaneous growth rate will be:

$$\frac{dN}{dt} = (0.005)(1100)\left[1-(1100/1000)\right]$$

$$= -0.55 \text{ fish}/\text{day}$$

The growth rate is negative because the additional stock pushes the population above its carrying capacity.

2.3. The equation for the death rate is linear; as in the simple logistic, the greater the number of turtles in the population, the greater the death rate. However, the equation for the birth rate is quadratic; it includes an N^2 term. This quadratic equation generates an Allee effect for reproduction: the birth rate first increases and then decreases with population size. Substitute different values of N into the birth and death functions to construct the graph shown on the next page.

Notice that the birth and death curves intersect in two different places. These points represent two different equilibrium population densities. One point of intersection is at a population size of approximately 34 turtles. If we are to the right of this equilibrium, the death rate exceeds the birth rate and the population declines, as shown by the arrow pointing to

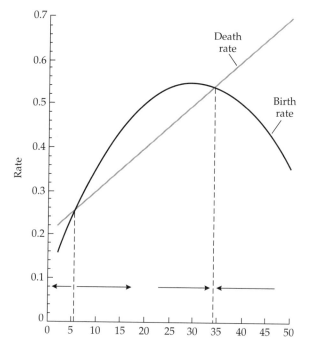

the left. If the population is less than 34, the birth rate exceeds the death rate and the population increases, as shown by the right-pointing arrow. Thus, this larger equilibrium point is **stable**.

The smaller equilibrium occurs at a population size of approximately 6 turtles. If the population is greater than 6, the birth rate exceeds the death rate, and the population will continue to increase until the equilibrium of 34 is reached. But if the population is less than 6, the death rate exceeds the birth rate, and the population declines to zero. Thus, this second equilibrium is **unstable**. By incorporating an Allee effect into the birth rate, we generate a minimum population size (6) that is necessary for the population to persist. This result is in contrast to the simple logistic model, in which the population always increased as long as it was below carrying capacity. See the Appendix for additional discussion of stable and unstable equilibrium points.

2.4. We wish to compare the growth rates of two populations, one of which is x individuals above carrying capacity, and one of which is x individuals below carrying capacity. For the first population, let $N = K + x$. Substituting into Equation 2.1 gives:

$$\frac{dN}{dt} = r(K+x)\left(1 - \frac{K+x}{K}\right)$$

For the population that is below carrying capacity, $N = K - x$, so its growth is represented by:

$$\frac{dN}{dt} = r(K-x)\left(1 - \frac{K-x}{K}\right)$$

In order to determine which growth rate is larger, we compare the size of both equations by factoring out equivalent terms:

$$r(K+x)\left(1 - \frac{K+x}{K}\right) \overset{?}{\longleftrightarrow} r(K-x)\left(1 - \frac{K-x}{K}\right)$$

Dividing through by r and substituting K/K for 1 gives:

$$(K+x)\left(\frac{K}{K} - \frac{K+x}{K}\right) \overset{?}{\longleftrightarrow} (K-x)\left(\frac{K}{K} - \frac{K-x}{K}\right)$$

After subtracting and dividing through by K, this simplifies to:

$$(K+x)(-x) \overset{?}{\longleftrightarrow} (K-x)(x)$$

Notice that the expression on the left is negative, because this is the growth rate when the population is above K. Because we are interested in the *magnitude* of growth, we take the absolute value of both sides of the inequality. Then, dividing through by x yields:

$$(K+x) > (K-x)$$

This result proves that the decline of a a population above its carrying capacity is always faster than the increase from below carrying capacity. For this reason, the average population size will always be less than the average carrying capacity in a variable environment.

2.5. If the birth rate is density-dependent and the death rate is density-independent then:

$$b' = b - aN$$
$$d' = d$$

Substituting these two terms back into Equation 2.1 gives:

$$\frac{dN}{dt} = (b - aN - d)N$$
$$\frac{dN}{dt} = [(b - d) - aN]N$$

Treating $(b - d)$ as r, we have:

$$\frac{dN}{dt} = rN\left(1 - \frac{aN}{r}\right)$$

Because a and r are both constants, we can define K as r/a, which again leads to the logistic equation:

$$\frac{dN}{dt} = rN\left(1 - \frac{N}{K}\right)$$

You can think of this as a special case of Expression 2.5, in which the constant c equals zero because the death rate is density-independent.

2.6. Because this is a slow-growing population with seasonal fluctuations in carrying capacity, we can use Equation 2.7, with a mean carrying capacity of 500 larvae, and an amplitude of 250 larvae. With these values, the average population size is predicted to be:

$$\sqrt{(500)^2 - (250)^2} = 433 \text{ larvae}$$

Because the population growth rate is slow, we expect this population to respond sluggishly to the seasonal changes in carrying capacity and not show much variation in population size.

CHAPTER 3

3.1. Here are the $l(x)$ schedules from Table 3.5, with the y axis plotted on a base 10 logarithmic scale:

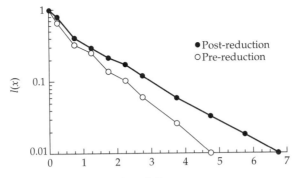

3.2a. Life-table data are calculated as follows:

x	$S(x)$	$b(x)$	$l(x) =$ $S(x)/S(0)$	$g(x) =$ $l(x+1)/l(x)$	$l(x)b(x)$	$l(x)b(x)x$	Initial estimate $e^{-rx}l(x)b(x)$	Corrected estimate $e^{-rx}l(x)b(x)$
0	500	0.0	1.00	0.8	0.00	0.00	0.000	0.000
1	400	2.5	0.80	0.1	2.00	2.00	0.965	0.946
2	40	3.0	0.08	0.0	0.24	0.48	0.056	0.054
3	0	0.0	0.00		0.00	0.00	0.000	0.000
			$R_0 =$ $\Sigma l(x)b(x)$	= 2.24 offspring	$\Sigma = 2.48$	$\Sigma = 1.021$	$\Sigma = 1.000$

$G = \dfrac{\Sigma l(x)b(x)x}{\Sigma l(x)b(x)}$	= 1.107 years
r (estimated) $= \ln(R_0)/G$	= 0.729 individuals/ (individual • year)
Correction added to estimated r	= 0.020
r (Euler)	= 0.749 individuals/ (individual • year)

3.2b. Stable age and reproductive value distributions are calculated as follows, using $r = 0.749$:

x	$l(x)$	$b(x)$	Stable age distribution		Reproductive value distribution			
			$l(x)e^{-rx}$	$c(x)$	$e^{rx}/l(x)$	$e^{-ry}l(y)b(y)$	• $e^{-ry}l(y)b(y)$	$v(x)$
0	1.00	0.0	1.000	0.716	1.000	0.000	1.000	1.000
1	0.80	2.5	0.378	0.271	2.644	0.946	1.000	0.143
2	0.08	3.0	0.018	0.013	55.909	0.054	0.054	0.000
			• = 1.396					

3.3 Age-specific survival probabilities and fertilities are calculated as follows:

x	i	$l(x)$	$b(x)$	$P_i =$ $l(i)/l(i-1)$	$F_i =$ $b(i)P_i$
0		1.00	0.0		
1	1	0.80	2.5	0.80	2.00
2	2	0.08	3.0	0.10	0.30
3	3	0.00	0.0	0.00	0.00

The resulting Leslie matrix is:

$$\mathbf{A} = \begin{bmatrix} 2.0 & 0.3 & 0 \\ 0.8 & 0 & 0 \\ 0 & 0.1 & 0 \end{bmatrix} \text{ and the initial population vector is: } \mathbf{n}(0) = \begin{bmatrix} 50 \\ 100 \\ 20 \end{bmatrix}$$

Using Equations 3.8 and 3.10, we have:

$n_1(1) = (2)(50) + (0.30)(100) + (0)(20) = 130$

$n_2(1) = (0.8)(50) = 40$

$n_3(1) = (0.1)(100) = 10$

$$\mathbf{n}(1) = \begin{bmatrix} 130 \\ 40 \\ 10 \end{bmatrix}$$

Repeating the calculation using this new vector gives:

$n_1(2) = (2)(130) + (0.30)(40) + (0)(10) = 272$

$n_2(2) = (0.8)(130) = 104$

$n_3(2) = (0.1)(40) = 4$

$$\mathbf{n}(2) = \begin{bmatrix} 272 \\ 104 \\ 4 \end{bmatrix}$$

CHAPTER 4

4.1a. Because the source of colonists is external and the extinctions are independent, the metapopulation is described by the island–mainland model (Equation 4.3). In this case, the equilibrium solution is found by solving Equation 4.4:

$$\hat{f} = \frac{p_i}{p_i + p_e} = \frac{0.2}{0.2 + 0.4} = 0.33$$

Approximately one island in three (33%) will support an ant lion population.

4.1b. Without the mainland population, colonization is strictly internal. In this case, the metapopulation is described by Equation 4.5 (internal colonization, independent extinctions), and the equilibrium solution is from Equation 4.6:

$$\hat{f} = (1 - \frac{p_e}{i}) = (1 - \frac{0.4}{0.2}) = -1$$

Because the equilibrium is less than zero, the island populations will all go extinct. Their persistence depends on the presence of the mainland population.

4.2. For the single pond, the probability of persistence is $(1 - 0.1) = 0.9$. For three ponds, we use Equation 4.2, with a new p_e of 0.50. In this case, the probability that a frog population will persist in at least one of the three ponds is:

$$P_x = 1 - (p_e)^x = 1 - (0.50)^3 = 0.875$$

So, in the short run, the probability of persistence is slightly higher for the population in a single pond (0.9) than for the subdivided populations in three ponds (0.875). In the long run, the best strategy will depend on the dynamics of the frog populations. If the subdivided populations quickly increase in size to 100 or more individuals, it might be worth the short-term risk to cultivate three viable populations rather than just one.

4.3. Because the metapopulation has a propagule rain and a rescue effect, its dynamics are described by Equation 4.7:

$$\frac{df}{dt} = (p_i)(1 - f) - ef(1 - f)$$
$$= (0.3)(1 - 0.4) - (0.5)(0.4)(1 - 0.4)$$
$$= 0.18 - 0.12$$
$$= 0.06 \text{ proportion of patches/time}$$

Because the "growth rate" is greater than zero, the metapopulation is increasing. Also, we could have solved for the equilibrium fraction of sites occupied with Equation 4.8:

$$\hat{f} = \frac{p_i}{e} = \frac{0.3}{0.5}$$
$$= 0.6 \text{ proportion of patches}$$

Because 40% of the population sites are occupied, $f = 0.4$. This is below the equilibrium value, also demonstrating that the metapopulation is expanding.

CHAPTER 5

5.1. Here is the plot of the state space, with the initial population sizes indicated by the star:

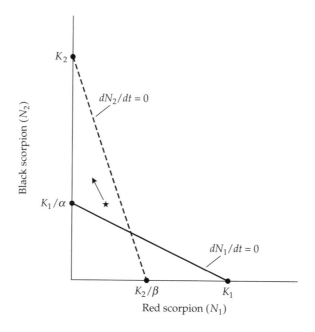

These isoclines define an unstable equilibrium. From the initial densities, the black scorpions will increase and the red scorpions will decrease in the short run. Eventually, the red scorpions will be driven to extinction, and the black scorpions will persist at their carrying capacity (K_2) of 150.

5.2. To answer this question, we use the inequalities in Table 5.1. Coexistence requires that:

$$\frac{1}{\beta} > \frac{K_1}{K_2} > \alpha$$

$$\frac{1}{0.5} > \frac{K_1}{100} > 1.5$$

$$2 > \frac{K_1}{100} > 1.5$$

A minimum carrying capacity of 151 individuals for species 1 is necessary to satisfy this inequality and ensure coexistence.

If species 1 is going to win in competition, then:

$$\frac{1}{\beta} < \frac{K_1}{K_2} > \alpha$$

$$2 < \frac{K_1}{100} > 1.5$$

For this to happen, the carrying capacity of species 1 must be greater than 200 individuals.

5.3. The following diagram illustrates the two isoclines for the initial stable equilibrium. The arrow indicates the shift in the "prey" isocline that allows the "predator" to win in competition:

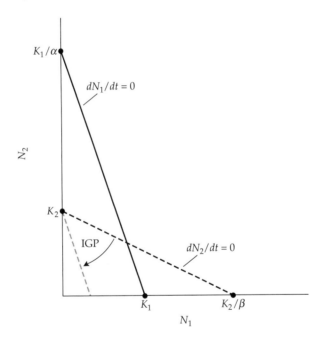

CHAPTER 6

6.1. From Equation 6.3, the solution for the victim isocline is $0.1/0.001 = 100$ spiders, and the solution for the predator isocline (Equation 6.4) is $0.5/0.001 = 500$ flies. Plotting the isoclines and the initial population sizes gives:

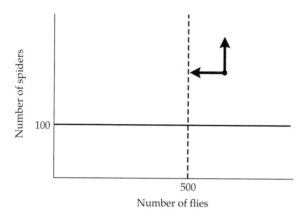

Because we are above the victim isocline, there are too many predators, so the fly population will decline. However, we are to the right of the predator isocline, so there are enough victims to allow the spider population to increase. Both populations will undergo smooth cycles.

6.2. From Equation 6.5, we first can solve for the missing value of q:

$$10 = \frac{2\pi}{\sqrt{0.5q}}$$

$$100 = \frac{4\pi^2}{0.5q}$$

$$50q = 4\pi^2$$

Thus, $q = 0.7896$. If q is doubled in size and r is 0.5, the period of the cycle becomes:

$$\text{period} = \frac{2\pi}{\sqrt{(0.5)(2)(0.7896)}} = \frac{6.283}{0.889} = 7.07 \text{ years}$$

So, the period of the cycle decreases to approximately 7 years.

6.3. The state-space drawing is as follows:

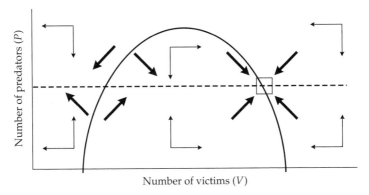

Number of victims (V)

From this drawing, it is easy to see that the equilibrium on the right is stable, and predator and prey will coexist. However, the equilibrium on the left is unstable, and if the victim population falls below a critical minimum, the predators will drive it to extinction. Note the similarity between this analysis and that of the Allee effect in the single-species logistic model (see Problem 2.3). However, do not confuse the state-space graphs with the single-species graphs of density-dependent birth and death rates!

6.4a. Because $k = 1/h$, we have:

$$D = \frac{1}{\alpha h}$$

$$\alpha D = \frac{1}{h}$$

$$\alpha D = k$$

$$\alpha = \frac{k}{D} = \frac{100}{5} = 20\,[\text{victims}/(\text{victim} \cdot \text{hour} \cdot \text{predator})]$$

6.4b. From Equation 6.8:

$$\frac{n}{t} = \frac{kV}{D+V}$$

$$= \frac{100(75)}{5+75} = \frac{7500}{80} = 93.8\,\text{prey/hour}$$

CHAPTER 7

7.1a. Plugging into Equation 7.1 gives $S = (8.759)(120)^{0.113} = 15.04$, which is close to the observed richness of 17 species.

b. With only half of the available area, the predicted species richness is $S = (8.759)(60)^{0.113} = 13.9$. So, roughly 14 species should be present, if the equation is completely accurate. Can you list several reasons why this forecast might be seriously incorrect?

7.2. The equilibrium depends on both the immigration and the extinction curves. Therefore, if the large island (A_1) is very isolated, it may contain fewer species than the small island (A_2):

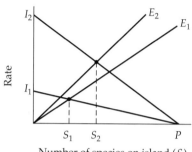

Number of species on island (S)

7.3. First, we need to use Equation 7.4 to solve for I. Equation 7.4 is written in terms of $\hat{S}$, but we can rearrange it to give a solution for I:

$$\hat{S} = \frac{IP}{I+E}$$

$$\hat{S}(I+E) = IP$$

$$\hat{S}E = IP - I\hat{S}$$

$$I = \frac{\hat{S}E}{P-S}$$

Plugging into this last expression yields $I = (75)(10)/(100 - 75) = 30$ species immigrations per year. If the immigration rate is doubled, $I = 60$. From Equation 7.4, the new species equilibrium is $(100)(60)/(60 + 10) = 85.7$ species. From Equation 7.5, the new turnover rate is $(60)(10)/(60 + 10) = 8.57$ species per year.

7.4. The calculation from Equation 7.6 is tedious, so we will illustrate it in detail only for Island 1. First, we must calculate x1, the relative area of Island 1. From Expression 7.9:

$$x_1 = \frac{a_1}{A} = \frac{110}{110 + 100 + 10 + 5} = 0.489$$

Now we can use Expression 7.13 to find the probability of occurrence for each of the six species. Remember that n_j in this expression is the total abundance of each species, which we obtain by summing the rows of the data matrix:

$Species\ A \quad 1-(1-0.489)^3 \quad = 0.867$

$Species\ B \quad 1-(1-0.489)^1 \quad = 0.489$

$Species\ C \quad 1-(1-0.489)^{10} = 0.999$

$Species\ D \quad 1-(1-0.489)^6 \quad = 0.982$

$Species\ E \quad 1-(1-0.489)^2 \quad = 0.739$

$Species\ F \quad 1-(1-0.489)^4 \quad = 0.932$

From Equation 7.6, the expected species richness on Island 1 is:

$$E(S_1) = 0.867 + 0.489 + 0.999 + 0.982 + 0.739 + 0.932 = 5.008$$

If we continue these calculations for all four islands, we obtain:

	Observed S	Expected S
Island 1	6	5.008
Island 2	1	4.835
Island 3	3	1.020
Island 4	3	0.540

The expected species richness does not match the observed very well. For example, the second largest island had only a single species on it, whereas the passive sampling model predicted almost 5 species. The two smallest islands each supported 3 species, but the passive sampling model predicted 1 or zero species. These data suggest that a random placement of individuals cannot account for the observed species richness data.

CHAPTER 8

8.1a. 0.030

8.1b. Algae (it has the highest transition to free space: $p_{ij} = 0.678$)

8.1c. Soft corals → Encrusting acroporid corals
    Soft corals → Algae
    Soft corals → Pocilloporid corals
    Algae → Pocilloporid corals
    Massive corals → Tabular *Acropora*
    Massive corals → Pocilloporid corals
    Pocilloporid corals → Tabular *Acropora*
    Pocilloporid corals → Soft corals
    Pocilloporid corals → Algae
    Pocilloporid corals → Massive corals

8.2. You can answer this question by using the matrix multiplication procedure described in this chapter. However, if you think about this particular set of numbers, you will see that you can be lazy. Just add up the transition elements for each row of the matrix, multiply by the total by 100, and you will have the number of patches for each stage.

Stage	Number of patches at $t = 1$
Encrusting acroporid coral	65.8
Tabular *Acropora*	37.5
Bushy *Acropora*	103.4
Staghorn *Acropora*	73.1
Soft corals	88.0
Algae	11.1
Massive corals	47.5
Pocilloporid corals	23.7
Free space	449.8

Why does this calculation work? Can you use this trick for other patch numbers?

If you were observant, you noticed that your patch numbers now sum to 899.90, rather than to 900.00, as they should. The discrepancy arises because the columns of the published matrix do not sum exactly to 1.00, probably because of rounding errors.

Glossary

Bracketed numbers refer to chapters.

α See competition coefficient. [5] See capture efficiency. [6]
αV See functional response. [6]
β See competition coefficient. [5] See conversion efficiency. [6]
βV See numerical response. [6]
λ See finite rate of increase. [1]
λ_s See immigration rate. [4]
μ_s See extinction rate. [4]

σ_x^2 See variance. [1]

σ_N^2 See variance in population size. [1]

σ_r^2 See variance in r. [1]

$1 - N/K$ See unused portion of the carrying capacity. [2]

A See transition matrix. [3, 8]

age class Individuals within a certain age interval are classified into a single age class. Individuals of age class i include those individuals between the ages of $i - 1$ and i. Thus, individuals in the first age class include newborns through individuals about to celebrate their first birthday. Age classes are counted beginning with 1, whereas ages are counted starting at 0. The age class of an individual is represented by a subscripted variable. Thus, f_6 represents individuals in the sixth age class. [3]

age The age of an individual is the amount of time that has passed since its birth. Thus, newborns are classified as age 0, not age 1. The age of an individual is represented by a variable in parentheses. Thus, $f(6)$ refers to individuals of age 6. [3]

Allee effect An increase in the instantaneous birth rate (b) or a decrease in the instantaneous death rate (d) as population size increases. In most populations, birth rates decrease and death rates increase as the population grows. These negative effects of crowding are the typical way that density dependence expresses itself in population growth. In some cases, however, population growth is actually enhanced by increasing population size. Increased population growth can occur if individuals in groups are able to hunt more efficiently, find mates, or defend themselves more effectively against predators, parasites, or diseases. All of these activities would cause the rate of population growth to increase as the size of the population increased. Eventually, however, we expect negative effects of crowding to emerge so that population growth rates turn downward as abundance increases further. Allee effects can cause simple models to exhibit more complex dynamics, such as a minimum sustainable population size (see Problem 2.3). The Allee effect is named after the ecologist Warder C. Allee (1885–1955), who popularized the idea in an influential textbook, *Principles of Animal Ecology* (Allee et al. 1949). [2]

allelopathy Chemical interference competition among plants effected through the release of toxic, often aromatic, chemicals in the soil. [5]

amplitude In a population cycle, the amplitude is the difference between the maximum population size and the population size at its midpoint. The units of amplitude are numbers of individuals. [2]

annual A plant that lives for only one season. [3]

area effect Species number increases with island area. [7]

B See birth rate. [1]

b See instantaneous birth rate. [1]

$b(x)$ See fecundity schedule. [3]

birth rate (B) The change in the number of births in a population measured over a short time interval. Its units are births/time. [1]

birth-flow model An age-structured model in which births occur continuously during the interval of a particular age class, in contrast to a birth-pulse model. [3]

birth-pulse model An age-structured model in which births are pulsed or concentrated, so that individuals give birth to all of their offspring on the day they enter a new age class. [3]

$c(x)$ See stable age distribution. [3]

capture efficiency (α) The effect of a predator on the per capita growth rate of the prey population:

$$\left(\frac{1}{V}\frac{dV}{dt}\right)\left(\frac{1}{P}\right)$$

The units of α are victims/(victim • time • predator). [6]

carrying capacity (K) The maximum number of individuals that can be supported in a population that is growing according to the logistic growth equation. This limit reflects the availability of space, food, and other resources in the environment. The units of carrying capacity are numbers of individuals. [2]

cellular automata A mathematical model in which patches are placed in a spatial grid, and the transition rules for a patch depend on the state of the surrounding patches. [8]

chaos A special pattern of apparently unpredictable fluctuations in population size that is generated by a model that is entirely deterministic. Chaotic fluctuations arise in discrete models of population growth that have strong density dependence and large intrinsic rates of increase. Chaotic dynamics are not caused by random variation in the environment, although chaotic populations are very sensitive to initial conditions. [2]

climax community The endpoint of the classic facilitation model, the climax community is diverse, self-replacing, and relatively stable. Disturbances can remove the climax community and re-start the successional sequence. [8]

closed population A closed population is one in which there is no immigration or emigration. Therefore, the only forces that can change the size of a closed population are births and deaths. [1]

cohort A group of individuals in a population, all born at the same time, that are followed until death. A cohort analysis allows us to directly measure the mortality rates of different ages and to construct the $l(x)$ schedule for the population. [3]

cohort life table See horizontal life table. [3]

cohort survival ($S(x)$) The number of individuals of a cohort that have survived to age x. By definition, $S(0)$ is the number of individuals in the original cohort. [3]

competition coefficient (α) In the Lotka-Volterra competition model, α is the per capita effect of species 2 on the population growth rate of species 1, measured relative to the effect of species 1. Since it is a scalar constant, α is a dimensionless number without units. [5]

competition coefficient (β) In the Lotka–Volterra competition model, β is the per capita effect of species 1 on the population growth rate of species 2, measured relative to the effect of species 2. Since it is a scalar constant, β is a dimensionless number without units. [5]

competitive interactions Species negatively affect each other's population growth rate and depress each other's population size. [5]

constraints Constraints are physiological, mechanical, or evolutionary limitations that prevent the evolution of certain life history traits. These constraints may represent adaptations that evolved for other reasons, but limit certain kinds of life history evolution. [3]

continuous differential equation An idealized equation for the population growth rate in which the time steps between consecutive measurements of the population are infinitely small. Representing population growth as a continuous differential equation allows us to use the rules of calculus and integration to solve the equation. [1]

continuous population growth A population that is growing continuously has births and deaths occuring steadily, so that the trajectory of population size resembles a perfectly smooth curve. [1]

conversion efficiency (β) The ability of predators to convert each prey item captured into additional per capita growth rate:

$$\left(\frac{1}{P} \frac{dP}{dt} \right) \left(\frac{1}{V} \right)$$

The units of β are predators/(predator • time • victim). [6]

D See death rate. [1] See half-saturation constant. [6]

d See instantaneous death rate. [1]

dN/dt See population growth rate. [1]

damped oscillations Oscillations in which the period of the fluctuations becomes smaller with time. Damped oscillations converge on a stable equilibrium point. [2]

Darlington's rule Named after the biogeographer Philip J. Darlington, the rule is that, on oceanic islands, a tenfold increase in island area is needed for each doubling of species number. [7]

death rate (D), (q) In the exponential growth model, the death rate D is the change in the number of deaths in a population measured over a short time interval. Its units are deaths/time. [1] In the Lotka–Volterra predation model, the death rate q

is the instantaneous death rate for the predator population in the absence of the victim population. q is identical to d, the instantaneous death rate in the exponential growth model. Its units are predators/(predator • time). [6]

delay differential equation A continuous differential equation that includes a time lag, so that current population growth depends on population size at some point in the past. [2]

demographic stochasticity Uncertainty due to variation in the sequence of births and deaths in a population. Even in a constant environment (no variation in r), discrete births and deaths can cause population numbers to vary unpredictably. Demographic stochasticity is analogous to genetic drift, in which allele frequencies in a population vary by chance. Demographic stochasticity is not important in large populations because this source of random variation tends to average itself out over the long run. But in small populations, demographic stochasticity can generate a substantial risk of extinction, even in an exponential growth model where birth rates exceed death rates. In contrast, small populations are never at risk of extinction in a deterministic model of exponential growth, as long as r is greater than zero. [1]

density-dependent model A model in which the instantaneous birth and death rates (b and d) are influenced by the density (or size) of the population. These models typically incorporate the idea that crowding leads to a reduction in births and an increase in deaths, providing an effective brake on population growth. [1]

density-independent Population processes that are not affected by the current density (or size) of the population. If instantaneous birth and death rates (b and d) are density-independent, a population will grow exponentially because these rates will not change no matter how large the population is. [2]

deterministic model A model in which the parameters are constant and do not vary unpredictably with time. In a deterministic model, there is no element of chance or uncertainty in the calculation. If the starting conditions are not altered, a deterministic model will always produce the same result. [1]

dimensionless number A constant that represents the ratio of two quantities. Because the units of the two quantities are identical, they cancel, so that the ratio has no units associated with it. Examples include the finite rate of increase (λ), which is a ratio of consecutive population sizes in an exponentially growing population, and the reproductive value ($v(x)$), which is the ratio of expected future production of offspring of individuals of age x to the number of individuals of age x. [1]

discrete difference equation A mathematical model in which time is measured in discrete steps, rather than continuously. In population ecology, discrete difference equations are used to model population size in the next time step (N_{t+1}) as a function of population size in the current time step (N_t). If the time step is infinitely small, then a discrete difference equation is equivalent to a continuous differential equation. [1]

discrete growth factor (r_d) The constant proportion by which a population increases each time step in a discrete model of exponential population growth. The units are individuals/(individual • time). If the time step is infinitely small, then r_d is equivalent to r, the instantaneous rate of increase in the continuous model of exponential population growth. r_d also equals $\lambda - 1$, where λ is the finite rate of increase. [1]

distance effect Species number decreases with distance or isolation from a source pool of colonists. [7]

doubling time The amount of time it takes for a population to double in size. If a population is growing exponentially, it will double in size after a constant amount of time, no matter how large it has become. The doubling time is $\ln(2)/r$, where ln is the natural logarithm, and r is the instantaneous rate of increase. [1]

E See extinction rate. [4] See maximum extinction rate. [7]

emigration Individuals leaving a population and traveling to another location. Emigration and death are the two ways that populations can decrease in size. [1]

environmental stochasticity Uncertainty due to variation in environmental conditions. The environment can be modeled as a series of unpredictably good and bad years for population growth. In an exponential growth model, this uncertainty is expressed in the mean and variance in r, the instantaneous rate of increase. In an exponential growth model with environmental stochasticity, a population is at risk of extinction if the variation in r is too large relative to the mean of r. In contrast, a population is never at risk of extinction in the deterministic model of exponential growth as long as r is greater than zero. [1]

equilibrium model of island biogeography Developed by Robert H. MacArthur (1930–1972) and Edward O. Wilson (b. 1929), the model describes island species richness as an equilibrium between ongoing colonization of new species and extinction of resident island species. [7]

ergodicity A mathematical property of some ecological models. A system is ergodic if its eventual behavior is independent of its initial state. [8]

Euler equation First derived by the Swiss mathematician Leonhard Euler (1707–1783), the Euler equation gives an exact solution for the instantaneous rate of increase (r) in terms of the fecundity ($b(x)$) and survivorship ($l(x)$) schedules for the population. The Euler equation is:

$$1 = \int_0^k e^{-rx} l(x) b(x)$$

The evolutionary biologist Ronald A. Fisher (1890–1962) used the Euler equation to derive an expression for the reproductive value of individuals of different ages. [3]

exploitation competition Competition that occurs because species use a shared resource that is in limited supply. [5]

exponential population growth A simple model of population growth in which the population growth rate (dN/dt) is the product of the current population size (N) and the instantaneous rate of increase (r). A population that is increasing exponentially has constant per capita birth (b) and death (d) rates and behaves like a savings account that grows with compound interest. Exponential growth implies no limit to population size and an accelerating rate of population growth. Although no population in nature ever shows exponential growth for very long, all populations have the potential to increase exponentially because each individual can leave more than one offspring in the next generation. The model of exponential population growth is the foundation for most of our modeling efforts in population and community ecology. [1]

extinction rate (E), (μ_s) In metapopulation models, the extinction rate E is the proportion of sites occupied by populations that go extinct per unit time. In the equilibrium model of island biogeography, the extinction rate μ_s is the number of resident species on the island going extinct per unit time. [4]

F_i See fertility. [3]

f See fraction of sites occupied. [4]

facilitation model A model of succession in which each group of species that enters a patch alters the environment in a way that facilitates the entry of successive sets of species. The endpoint of the classic facilitation model is a self-replacing climax community. [8]

fecundity schedule ($b(x)$) The fecundity schedule gives the average number of offspring born per unit time to individual females of age x. Fecundities are always represented by non-negative real numbers. Fecundities of some ages may be zero if females of those ages are either pre- or post-reproductive. [3]

feeding rate (n/t) The rate at which individual predators capture prey. Its units are victims/(time • predator). [6]

fertility (F_i) The number of female offspring produced by females of age class i. Fertility coefficients are represented as entries in the first row of the Leslie matrix. Fertility can be calculated as $F_i = b(i)P_i$. Thus, the fecundity for females of age i is discounted by P_i, the survival of individuals through age class i. This discounting is necessary because females must survive through the age class in order to reproduce and have their offspring counted. This formula applies only to birth-pulse populations with a post-breeding census. [3]

finite rate of increase (λ) A ratio measuring the proportional change in population size from one time step to the next in a discrete model of exponential population growth. In a population that is increasing exponentially, $\lambda = N_{t+1}/N_t$, the ratio of population sizes in two consecutive time steps. Because it is a ratio, λ is a dimensionless number without units. λ is always a number greater than 0 because it is a ratio of two population sizes, which are positive numbers. λ is equal to $1.0 + r_d$, the discrete growth factor. If the time step is infinitely small, then λ is also equal to e^r, where e is a constant, the base of the natural logarithm ($e \approx 2.718$), and r is the instantaneous rate of increase. [1]

first-order Markov model A Markov model in which transition elements depend only on the current state of the assemblage. See also homogeneity. [8]

fraction of sites occupied (f) The fraction of sites occupied is the proportion of available sites that contain populations. It is a number that varies between a minimum of 0.0 (regional extinction) and a maximum of 1.0 (landscape saturation). [4]

functional response (αV) The rate of victim capture by a predator as a function of victim abundance. Its units are victims/(predator • time). [6]

G See generation time. [3]

$g(x)$ See survival probability. [3]

generation time (G) The generation time is an estimate of the amount of time it takes one cohort to grow up and replace another. One measure is the average age of the parents of all the offspring produced by a single cohort. It can be calculated from the survivorship and fecundity schedules as:

$$G = \sum l(x)b(x)x / \sum l(x)b(x)$$

It is measured in units of time. [3]

h See handling time. [6]

half-saturation constant (D) In a Type II and Type III functional response, D is the victim abundance at which the predator feeding rate is half the maximum (k). Its units are numbers of victims. [6]

handling time (h) The amount of time per prey item a predator requires to capture and eat a victim. [6]

homogeneity A mathematical property of some Markov models. A transition matrix is homogenous if its elements do not change through time. [8]

horizontal life table A life table for which the survivorship schedule ($l(x)$) is calculated by directly following a cohort of individuals from birth to death. [3]

I See immigration rate. [4] See maximum immigration rate. [7]

IGP See intraguild predation. [5]

immigration Individuals entering a population from another location. Immigration and birth are the two ways that populations can increase in size. [1]

immigration rate (I), (λ_s) In metapopulation models, the immigration rate I is the proportion of sites successfully colonized per unit time. In the equilibrium model of island biogeography, the immigration rate λ_s is the number of new species arriving on an island per unit time. "New" species are those that occur in the source pool but do not presently occur on the island. [4]

individual-based model An ecological model in which a computer is used to simulate the birth, growth, dispersal, and death of individuals in a population. Such models are realistic but computationally intensive. [8]

inhibition model A model of succession in which each community or species that holds space inhibits other species from entering the assemblage. Succession occurs only when a disturbance removes the resident species, allowing new species to enter the community. [8]

instantaneous birth rate (b) The per capita birth rate of the population, which is the number of births per individual measured over a short time interval. It can be calculated by dividing the birth rate (B) by the current population size (N). Its units are births/(individual • time). [1]

instantaneous death rate (d) The per capita death rate of the population, which is the number of deaths per individual measured over a short time interval. It can be calculated by dividing the death rate (D) by the current population size (N). Its units are deaths/(individual • time). [1]

instantaneous rate of increase (r) The instantaneous rate of increase is $b - d$, the difference between the instantaneous birth rate (b) and the instantaneous death rate (d). In a simple model of exponential population growth, the instantaneous rate of increase is also equal to the per capita rate of population increase:

$$\left(\frac{1}{N} \frac{dN}{dt} \right)$$

The units of r are individuals/(individual • time). [1]

interference competition Competition in which individuals behave in a way that reduces the exploitation efficiency of a competitor. Examples include territoriality, behavioral interference among animals, and allelopathic interactions among plants. [5]

internal colonization A metapopulation model in which propagules only originate from occupied sites. Thus, colonization ceases if there is regional extinction because there is no external source of propagules. [4]

interspecific competition Competition among individuals of different species for limiting resources. [5]

intraguild predation (IGP) Competitors that exploit common, limiting resources but also interact with one another as predator and prey. IGP is common in nature, and it can either reverse or reinforce the outcome of competitive interactions between species. [5]

intraspecific competition Competition among individuals of the same species for limiting resources. The logistic growth model (Equation 2.1) incorporates the idea of intraspecific competition. [5]

intrinsic rate of increase See instantaneous rate of increase. [1]

island–mainland model A metapopulation model in which local extinctions are independent of one another and colonization occurs via a propagule rain. At the community level, this model is equivalent to the equilibrium model of island biogeography. [4]

iteroparous Iteroparous organisms are those that reproduce at more than one age in their life history. The fecundity schedule for an iteroparous organism would contain two or more non-zero entries for the reproductive ages. [3]

K See carrying capacity. [2]

k See maximum feeding rate. [6]

Leslie matrix A matrix representation of the birth and death parameters in an age-structured population growth model. The entries in the first row of the matrix represent the fertilities of each age class, and the subdiagonals represent the survival probabilities from one age class to the next. The Leslie matrix was developed by the population ecologist Patrick H. Leslie.

$l(x)$ See survivorship schedule. [3]

local extinction The disappearance of a single population within a metapopulation. [4]

logistic growth model A model of population growth that incorporates the concept of resource limitation and density dependence in the instantaneous birth rate and/or the instantaneous death rate. The logistic growth model was introduced to ecology by Pierre F. Verhulst (1804–1849). It generates a characteristic S-shaped curve in which the rate of population growth first accelerates, then decelerates. The population will achieve a constant carrying capacity K, which reflects the resources available in the environment. If the population begins above its carrying capacity, it will decrease in size until K is achieved. The exponential growth model can be derived as a special case of the logistic growth model in which the instantaneous birth and death rates are both density-independent. The formula for logistic population growth is $dN/dt = rN(1 - N/K)$. [2]

Malthusian parameter See instantaneous rate of increase. The term refers to the Reverend Thomas R. Malthus (1766–1834). Malthus' famous "Essay on the

Principle of Population" (1798) discussed the implications of exponential population growth for humans. [1]

Markov model A matrix model of succession or population growth in which a transition matrix is sequentially multiplied by a vector to produce change in the stages of a population or community. [3, 8]

maximum extinction rate (E) The maximum rate at which resident species on an island go extinct. This maximum is reached when all of the source pool species occur on the island ($S = P$). [7]

maximum feeding rate (k) In a Type II and Type III functional response, the maximum feeding rate k is the asymptotic per capita feeding rate of predators. Its units are victims / (time • predator). [6]

maximum immigration rate (I) The maximum rate at which new species arrive on an island. This maximum occurs when the island has no resident species. [7]

mean ($\bar{x}$) The arithmetic mean is the average, or central tendency, of a distribution or series of numbers. It is calculated as $\sum x/n$, where n is the sample size. In other words, the values are added up and then divided by the number of observations in the sample. The mean is the most natural measure of central tendency, but it is by no means the only one. Other measures include the median (the midpoint in the series of ranked observations) and the mode (the most common value in the sample). Other means, such as the geometric and harmonic means, can also be calculated. Both of these measures are useful in certain ecological analyses because they reflect the multiplicative nature of population growth. [1]

mean population size ($\bar{N}$) The central tendency for population size. In a stochastic growth model, the mean population size would be the average N generated by many runs of the model. [1]

mean r ($\bar{r}$) The average instantaneous rate of increase. This average reflects the central tendency of the per capita growth rate in a variable environment. In models of environmental stochasticity, the mean r is used to predict the mean and variance in population size. [1]

metapopulation A group of several local populations that are linked by immigration and emigration. The movement of individuals between populations affects local dynamics, so that the growth and persistence of the individual patches in the metapopulation is different from the growth and persistence of an isolated population. [4]

monocarpic A plant that reproduces at only one age. See semelparous. [3]

N See size of the population. [1]

$\bar{N}$ See mean population size. [1]

n/t See feeding rate. [6]

net reproductive rate (R_0) The mean number of female offspring produced by a female over her lifetime. The net reproductive rate is the gross number offspring produced, discounted by the chances of female survivorship through different ages. Its units are numbers of offspring, and it is calculated from the fecundity and survivorship schedules as $R_0 = \sum l(x)b(x)$. [3]

neutral equilibrium An equilibrium that remains in place until the system is perturbed, and it comes to rest at a new equilibrium. The simple Lotka–Volterra predation model (Equations 6.1 and 6.2) is an example of a neutral equilibrium. The predator and victim populations undergo population cycles whose amplitude is determined by the initial population sizes. If the populations are perturbed from these cycles, they will begin cycling with a new amplitude that is determined by the new starting conditions. A physical analogy of a neutral equilibrium is a marble resting on a flat surface. If the marble is displaced, it will come to rest in a new location and remain there unless it is displaced again. [4]

non-overlapping generations A population in which parents die before their offspring are born. A population with non-overlapping generations usually will not have any age structure and can be modeled with a discrete difference equation. [1]

numerical response (βV) The per capita growth rate of the predator population as a function of victim abundance. Its units are $(1/P)(dP/dt)$. [6]

optimal yield The harvesting level for a renewable resource that will sustain the resource and generate the maximum long-term yield. Unfortunately, the optimal yield for long-term harvesting is usually less than the optimal yield for short-term profits. As a consequence, most fisheries stocks and other renewable resources are chronically over-harvested. [2]

p_e See probability of local extinction. [4]

p_i See probability of local colonization. [4]

P See source pool. [7]

P_i See survival probability. [3]

P_n See probability of persistence. [4]

P_x See probability of regional persistence. [4]

paradox of enrichment A predator prey model that predicts that increasing the carrying capacity of the victim population destabilizes co-existence with the predator. The paradox of enrichment requires a hump-shaped victim isocline and a vertical predator isocline to generate the destabilization with increasing victim carrying capacity. [6]

passive sampling model A statistical model for the species-area relationship that does not invoke habitat diversity or extinction on small islands. The model treats islands as passive targets that differ in area and assumes that individuals of each species are distributed randomly throughout an archipelago. The model can be used to generate the expected species-area relationship in the absence of any biological forces. [7]

per capita Per individual. Per capita rates can be found by dividing the quantity by the number of individuals in the population (N), or, equivalently, multiplying by $1/N$. [1]

perennial A plant that lives for two or more seasons. [3]

period The period of a population cycle is the amount of time it takes for the population to go through one complete cycle and return to its current population size. The period of a cycle is measured in units of time. [2]

pioneer species The initital species to appear in an environment after a disturbance. Often these species have a number of r-selected life history traits that allow them to tolerate the harsh physical conditions that are present after a disturbance. [8]

polycarpic A plant that reproduces at more than one age. See iteroparous. [3]

population A population is a group of individuals, all of the same species, that live in the same place. Although it is sometimes difficult to define the physical boundaries of a population, the individuals within a population have the potential to reproduce with one another during the course of their lifetimes. [1]

population growth rate (dN/dt) The "velocity" or rate of change of population size. It is measured as the change in population size (dN) during a very small interval of time (dt). The units of population growth rate are individuals/time. [1]

post-breeding census A census of a population performed each year (or time-step) after all individuals have bred. [3]

pre-emptive competition Competition for attachment or rooting space as a limiting resource. Pre-emptive competition is common among plants, sessile marine invertebrates, and algae. [5]

primary succession Succession that proceeds on new substrate that has never previously been colonized by living organisms. Such succession may be slow because of the time required to build up a soil profile or nutrient levels necessary to sustain an entire community. [8]

principle of competitive exclusion The principle of competitive exclusion states that "complete competitors cannot coexist." In other words, there must be some difference between species in their resource utilization, which would be expressed in the competition coefficients α and β, for coexistence to occur. Note that the principle implies that resources are limiting. If resources are not limiting, because of predators or chronic disturbances that reduce the abundance of the competitors, two species can happily coexist using identical resources. [5]

probability of local colonization (p_i) The chance that an unoccupied site in a metapopulation is colonized during a given time period. [4]

probability of local extinction (p_e) The chance that a single local population will go extinct during a given time period. This probability will depend on the population size and population growth rate, as well as attributes of the site, such as its area and the resources it contains. [4]

probability of persistence (P_n) The chance that a single patch persists for n time periods, where the probability of extinction is p_e during each time period. The probability of persistence is $P_n = (1 - p_e)^n$, where $(1 - p_e)$ is the probability that extinction does *not* occur. [4]

probability of regional persistence (P_x) The chance that, out of a set of x patches, at least one of them will persist to the end of a given time period. If the patches are all identical, and they each have the same probability of local extinction p_e, the probability of regional persistence is $P_x = 1 - (p_e)^x$.

propagule rain A metapopulation model in which the probability of colonization does not depend on the fraction of sites occupied in the metapopulation. A propagule rain can result if there is a large, external source of colonists, or perhaps a long-lived seed bank that acts as a constant source of new propagules. [4]

proportional area (x_i) The relative area of an island, calculated by dividing the island's area by the summed area of all the islands in the archipelago. In the passive sampling model, the proportional area represents the probability that a randomly distributed propagule will land on a particular island. [7]

q See death rate. [6]

R_0 See net reproductive rate. [3]
r See instantaneous rate of increase. [1]
r_d See discrete growth factor. [1]
$\bar{r}$ See mean r. [1]

r–K selection A once-popular theory, r-K selection assumed that population density was the major selective force that determined the life history traits of an organism. Populations that were permanently maintained at low density (presumably because of external forces of mortality or disturbance) were labeled r-selected, and evolution was thought to favor early, semelparous reproduction, large r, many offspring with poor survivorship, a Type III survivorship curve, and small adult body size. Populations that experienced chronic high densities were labeled K-selected, and evolution was thought to favor late, iteroparous reproduction, small r, few offspring with good survivorship, a Type I survivorship curve, and large adult body size. [3]

recursion equation A discrete growth equation in which the "output" from one time step (N_{t+1}) forms the input for the next time step (N_t). Recursion equations can always be solved iteratively, by plugging in numbers consecutively until the desired time step is reached. For some recursion equations, there may be a mathematical solution that will produce N_t from a single calculation. [1]

regional extinction The disappearance of all local populations of a metapopulation. [4]

reproductive value ($v(x)$) Reproductive value is the expected number of offspring that remain to be born to individuals of age x, relative to the number of individuals of age x. By definition, the reproductive value of newborns ($v(0)$) is 1.0. Reproductive value for individuals of age x can be calculated as:

$$v(x) = \frac{e^{rx}}{l(x)} \sum_{y=x+1}^{k} e^{-ry}l(y)b(y)$$

Note that the counting subscript in the summation sign has been increased by 1. This change is necessary for using the formula with real (discrete) data. [3]

rescue effect In metapopulation models, the rescue effect is the reduction in the probability of local extinction with an increase in the fraction of sites occupied. When more sites are occupied, there are more migrants entering a local population, and the boost in population size reduces the probability of local extinction. In the equilibrium model of island biogeography, the rescue effect is the reduction in the species extinction rate on near versus far islands. [4]

robust A model is robust if we can violate some of its assumptions and find that its predictions still hold up. Whether they are stated explicitly or not, all models, even verbal ones, imply a list of underlying assumptions. In some cases, the predictions of the model are very sensitive to these assumptions. For example, the exponential growth model depends critically on the assumption of constant per capita birth and death rates. Other assumptions, such as the absence of migration or time lags, are less critical to the prediction of exponential population increase. New ecological models can often be built by systematically violating or altering the assumptions of an existing model. [1]

s See stage vector. [8]

S(x) See cohort survival. [3]

secondary succession Succession that proceeds after an established community has been removed from a patch by a disturbance. Some elements of the community may re-establish themselves from within the patch, while other elements colonize from adjacent patches that were not disturbed. [8]

second-order Markov model A Markov model in which transition elements depend both on the current state of the assemblage and on the state in the previous time step. These models incorporate some aspects of the historical sequence of community development. [8]

seed bank An accumulation in the soil of long-lived seeds, which potentially can sprout many years after they were sown. A seed bank can complicate population dynamics by introducing a time lag into population growth. [1]

semelparous Semelparous or "big bang" reproduction is a life history strategy in which all reproduction is concentrated in a single age. The fecundity schedule for a semelparous organism would have zeroes for all ages except the single reproductive age. [3]

semi-Markov model A Markov model in which transition probabilities depend on the absolute amount of time that a patch is in a particular state. These models describe species with different life histories and persistence times. [8]

senescence Aging and physiological deterioration of individuals in post-reproductive ages. [3]

sink populations Populations for which the local birth rate is less than the local death rate and the immigration rate is greater than zero. Sink populations cannot persist in isolation because the birth rate does not exceed the death rate. Sink populations are net "importers" of individuals, and they depend on external immigration for their persistence. [4]

size of the population (N) The number of individuals in a population. [1]

source pool (P) The number of species in a mainland or source area that can potentially colonize an island. [7]

source populations Populations for which the local birth rate exceeds the local death rate, and the emigration rate is greater than zero. Source populations are net "exporters" of individuals. [4]

species–area relationship The non-linear increase in species richness as island area is increased. The relationship holds for most groups of species on many different kinds of "islands." [7]

species–area slope (z) On a log(species) versus log(area) plot, z is the slope of the fitted line. z is also the exponent in the power function $S = cA^z$, where S is the number of species, A is the area of the island, and z and c are fitted constants. [7]

stable age distribution ($c(x)$) The relative proportion of individuals represented in each age of an exponentially increasing (or decreasing) population. The formula for the stable age distribution is:

$$c(x) = \frac{e^{-rx}l(x)}{\displaystyle\sum_{x=0}^{k} e^{-rx}l(x)}$$

Once a population reaches a stable age distribution, these proportions remain constant through time. A stable age distribution implies that a population is growing with fixed survivorship ($l(x)$) and fecundity ($b(x)$) schedules. [3]

stable equilibrium An equilibrium is stable if a population always returns to it after a small perturbation. If individuals are added to the population at equilibrium, the population will decline until it returns to the equilibrium point. If individuals are removed from the population, it will increase until the equilibrium point is reached. A physical analogy of a stable equilibrium is a marble resting on the bottom of a smooth bowl. If the marble is displaced, it will always come to rest at the bottom of the bowl. [2]

stable limit cycle A population cycle that is stable. In other words, if the population size is perturbed, it will return to a pattern of cycles with the same amplitude and period. A stable limit cycle is not a single equilibrium point, but it nevertheless represents a stable equilibrium because a population will always return to it after a small perturbation. [2]

stage vector A vector of length n whose elements are the number of patches that exist in a particular state in a succession model. At equilibrium, this stage vector represents either the relative number of patches in each stage, or the relative amount of time that a single patch spends in each stage. [8]

state-space graph A state-space graph is one in which the abundances of two interacting species are plotted on the x and y axes. Thus, each point in the state-space graph represents a combination of abundances, and points along the axes represent combinations in which one of the species is missing. The state-space graph is used to plot species' isoclines in predation and competition models to illustrate the equilibrium points and trajectories of each species. [5]

static life table See vertical life table. [3]

stationary age distribution A stationary age distribution is one in which both the absolute and relative numbers of individuals represented in each age remain constant. A stationary age distribution is a special case of the stable age distribution in which r, the instantaneous rate of increase, equals 0.0. [3]

stochastic model A model in which some of the parameters vary unpredictably with time. Stochastic models reflect random or chance events in nature, or complex, changing phenomena that are too complicated to model directly. In a stochastic model, the population track will reflect an element of chance or uncertainty. Consequently, if the model is run twice with the same starting conditions, it will generate somewhat different answers each time. Although each run of a stochastic model is unique, if the model is run many times, there is usually an expected mean and variance for the predicted population size. [1]

succession Change in community structure through time. [8]

survival probability ($g(x)$), P_i The survival probability $g(x)$ for individuals of age x is defined as the probability that an individual alive at age x will be alive at age $x + 1$. It is calculated from the survivorship schedule as $g(x) = l(x + 1)/l(x)$. The survival probability P_i for individuals in age class i is defined as the probability that an individual that is alive in age class i will survive to age class $i + 1$. It is calculated from the survivorship schedule as $P_i = l(i)/l(i - 1)$. This formula is applicable only to birth-pulse populations with a post-breeding census. Survival probabilities appear as subdiagonal elements in the Leslie matrix. [3]

survivorship schedule ($l(x)$) The survivorship schedule gives the probability that an individual survives from birth to the start of age x. By definition, the survival of newborns is 100%, so $l(0)$ always equals 1.0, and the survival of the last age class is 0.0. Between these endpoints, survivorship decreases (or stays the same) with each consecutive age class. [3]

T See turnover rate. [7]

target effect The effect of island area on the immigration rate. In the original equilibrium model, island area affects only the extinction rate. However, if large islands intercept more propagules, they will have both lower extinction rates and higher immigration rates than small islands. [7]

time lag A delay in the response of a population to factors controlling population growth. If the time lag is zero, the growth of a population depends on its current population size, as in a continuous differential equation. However, if the time lag is, say, 5 years, population growth will depend not on the current size of the population, but on the size of the population 5 years ago. Time lags can generate complex population dynamics, including damped oscillations and stable limit cycles. [2]

tolerance model A succession model in which species neither facilitate nor inhibit later species from entering the community. The tolerance model can serve as a kind of null hypothesis for testing the inhibition and facilitation succession models. [8]

tradeoffs In life history theory, tradeoffs occur when large values for one trait lead to small values for another. For example, individuals that invest heavily in early reproduction may have poor survivorship in later life. Tradeoffs arise, in part, because organisms have a limited amount of energy that must be apportioned between functions that promote growth, survivorship, and reproduction. [3]

transition matrix A square $n \times n$ matrix whose elements p_{ij} give the probability of a transition from state j to state i in a single time step. In successional models, these probabilities describe changes between community states, whereas in age-structured demographic models, these probabilities describe changes between different ages or life history stages. In demographic models, the elements of the transition matrix may be larger than 1.0 if they represent reproductive transitions. Otherwise, the elements of the matrix are bounded by 0.0 and 1.0. [3, 8]

turnover rate (*T*) The number of species arriving or disappearing per unit time for an island community in equilibrium. [7]

Type I functional response A linear increase in the number of prey consumed per predator per unit time as victim abundance increases. The slope of the line is α, the capture efficiency. A Type I functional response is built into the simple Lotka–Volterra predator–prey equations and tends to stabilize predator–prey dynamics. It is unrealistic because it assumes that predators can always increase their feeding rates. [6]

Type I survivorship curve A survivorship curve in which survival probabilities are relatively high for young individuals and relatively low for old individuals. A Type I survivorship curve characterizes many mammals (including humans) that invest heavily in parental care of offspring. [3]

Type II functional response An asymptotic curve describing the number of prey consumed per predator per unit time as victim abundance increases. The shape of the curve is determined by the maximum feeding rate (k) and the half-saturation constant (D). A Type II functional response can arise because of predator satiation and constraints on handling time. It is generally destabilizing because it requires more predators to control the victim population as victim abundance increases. [6]

Type II survivorship curve A survivorship curve in which survival probabilities are relatively constant across different ages. Relatively few species show true Type II survivorship curves. See Problem 3.1 for an example. [3]

Type III functional response An asymptotic curve describing the number of prey consumed per predator per unit time as victim abundance increases. The curve is S-shaped, so the feeding rate accelerates at low victim abundances, but then decelerates and approaches an asymptote at high victim abundances. A Type III functional response can arise because of search images and predator switching behavior. It is generally stabilizing at low victim abundances, but destabilizing at high victim abundances. [6]

Type III survivorship curve A survivorship curve in which survival probabilities are relatively low for young individuals and relatively high for old individuals. A Type III survivorship curve characterizes many plants and invertebrates that produce large numbers of offspring, few of which survive. [3]

unstable equilibrium An equilibrium that is not returned to if the populations are perturbed. In an unstable equilibrium, the populations will not return to their equilibrium values if they are perturbed. Instead, they will travel towards a different, more stable, equilibrium. A physical analogy of an unstable equilibrium is a marble resting on the top of a smooth, inverted bowl. If the marble is bumped slightly in any direction, it will roll off its peak and come to rest somewhere else. [5]

unused portion of the carrying capacity (1 – N/K) In the logistic growth model, this term represents the portion of the carrying capacity that has not already been used up by the population at its current size (N). This term causes population growth to be fastest when the population is close to zero, slowest when it is close to carrying capacity, and negative when the population has exceeded carrying capacity. [2]

$v(x)$ See reproductive value. [3]

variance $\left(\sigma^2_x\right)$ The variance is a measure of the spread or uncertainty about the mean. It is calculated as:

$$\sum (x - \bar{x})^2 / (n - 1)$$

where $\bar{x}$ is the mean and n is the sample size. In other words, the mean is subtracted from each observation x, and the result is squared. These squared values are summed, and the total is divided by the sample size minus one. The variance quantifies the extent to which each observation tends to differ from the overall mean. If all of the observations are the same, then they do not differ from the mean, and the variance is zero. The more unpredictable the observations are, the larger the variance. Note that the mean and the variance quantify two distinct aspects of a series of numbers. The mean measures the central tendency of the observations, whereas the variance measures the extent to which those observations differ from the central tendency. A population with a large mean might have a small variance, and vice versa. [1]

variance in population size $\left(\sigma_N^2\right)$ The variability or uncertainty in mean population size. In a stochastic growth model, the variance in N would reflect the variability in population size generated by many runs of the model. In many stochastic models, the variance in N increases as the length of the time series is extended. If the variance in N is too great, populations may be at risk of extinction. [1]

variance in r $\left(\sigma_r^2\right)$ The variance in the instantaneous rate of increase. This variance measures the variability of r, reflecting good and bad times for population growth. In models of environmental stochasticity, a population is at risk of extinction if the variance in r is too large compared to the mean r. [1]

vertical life table A life table for which the survival probabilities ($g(x)$) are estimated indirectly by comparing the relative sizes of consecutive age classes. This analysis assumes that the population has achieved a stationary age distribution. [3]

$\bar{x}$ See mean. [1]

x_i See proportional area. [7]

z See species–area slope. [7]

Literature Cited

Abramsky, Z., M. L. Rosenzweig and B. Pinshow. 1991. The shape of a gerbil isocline measured using principles of optimal habitat selection. *Ecology* 72: 329–340. [5]

Abramsky, Z., O. Ovadia and M. L. Rosenzweig. 1994. The shape of a *Gerbillus pyramidum* (Rodentia: Gerbillinae) isocline: an experimental field study. *Oikos* 69: 318–326. [5]

Açkakaya, H. R. 1992. Population cycles of mammals: evidence for a ratio-dependent predation hypothesis. *Ecological Monographs* 62: 119–142. [6]

Allee, W. C., A. E. Emerson, O. Park, T. Park and K. P. Schmidt. 1949. *Principles of Animal Ecology*. W. B. Saunders, Philadelphia. [2]

Anderson, R. M. and R. M. May. 1978. Regulation and stability of host–parasite population interactions. I. Regulatory processes. *Journal of Animal Ecology* 47: 219–249. [6]

Arcese, P. and J. N. M. Smith. 1988. Effects of population density and supplemental food on reproduction in the song sparrow. *Journal of Animal Ecology* 57: 119–136. [2]

Arditi, R. and L. R. Ginzburg. 1989. Coupling in predator-prey dynamics: ratio-dependence. *Journal of Theoretical Biology* 139: 311–326. [6]

Arrhenius, O. 1921. Species and area. *Journal of Ecology* 9: 95–99. [7]

Berryman, A. 1992. The origins and evolution of predator–prey theory. *Ecology* 73: 1530–1535. [6]

Boerema, L. K. and J. A. Gulland. 1973. Stock assessment of the Peruvian anchovy (*Engraulis ringens*) and management of the fishery. *Journal of the Fisheries Research Board of Canada* 30: 2226–2235. [2]

Botkin, D. B. 1992. *Forest Dynamics: An Ecological Model*. Oxford University Press, Oxford. [8]

Brown, J. H. and A. Kodric-Brown. 1977. Turnover rates in insular biogeography: effect of immigration on extinction. *Ecology* 58: 445–449. [7]

Caswell, H. 2001. *Matrix Population Models*, 2nd edition. Sinauer Associates, Sunderland, Mass. [3,8]

Caughley, G. 1977. *Analysis of Vertebrate Populations*. Wiley, New York. [3]

Clements, F. E. 1904. The development and structure of vegetation. *Bot. Surv. Nebraska* 7: 5–175. [8]

Clements, F. E. 1936. Nature and structure of the climax. *Journal of Ecology* 24: 252–284. [8]

Coleman, B. D., M. A. Mares, M. R. Willig and Y.-H. Hsieh. 1982. Randomness, area, and species richness. *Ecology* 63: 1121–1133. [7]

Connell, J. H., and R. O. Slatyer. 1977. Mechanisms of succession in natural communities and their role in community stability and organization. *The American Naturalist* 111: 1119–1144. [8]

Connell, J. H., I. R. Noble, and R. O. Slatyer. 1987. On the mechanisms of producing successional change. *Oikos* 50: 136–137. [8]

Darlington, P. J. 1957. *Zoogeography: The Geographical Distribution of Animals.* Wiley, New York. [7]

den Boer, P. J. 1981. On the survival of populations in a heterogeneous and variable environment. *Oecologia* 50: 39–53. [4]

Dennis, B., P. L. Munholland and J. M. Scott. 1991. Estimation of growth and extinction parameters for endangered species. *Ecological Monographs* 61: 115–144. [1]

Diamond, J. 1986. Overview: Laboratory experiments, field experiments, and natural experiments. *In* J. Diamond and T. J. Case (eds.), *Community Ecology*, pp. 3–22. Harper & Row, New York. [8]

Diamond, J. M. 1972. Geographic kinetics: estimation of relaxation times for avifaunas of southwest Pacific islands. *Proceedings of the National Academy of Sciences, USA* 69: 3199–3203. [7]

Doak, D. F. and W. Morris. 1999. Detecting population-level consequences of ongoing environmental change without long-term monitoring. *Ecology* 80: 1537–1551. [8]

Dobson, A. P. and P. J. Hudson. 1992. Regulation and stability of a free-living host–parasite system: *Trichostrongylus tenuis* in red grouse. II. Population models. *Journal of Animal Ecology* 61: 487–498. [6]

Donovan, T. M. and C. Welden. 2001. *Exercises in Ecology, Evolution, and Behavior: Programming Population Models and Simulations with Spreadsheets.* Sinauer Associates, Sunderland, Mass. [Preface, 8]

Durrett, R. and S. A. Levin. 1994. Stochastic spatial models: a user's guide to ecological applications. *Philosophical Transactions of the Royal Society of London B* 343: 329–350. [8]

Elton, C. and M. Nicholson. 1942. The ten-year cycle in numbers of the lynx in Canada. *Journal of Animal Ecology* 11: 215–244. [6]

Ehrlich, P. R., R. R. White, M. C. Singer, S. W. McKechnie and L. E. Gilbert. 1975. Checkerspot butterflies: a historical perspective. *Science* 188: 221–228. [4]

Facelli, J. M. and S. T. A. Pickett. 1990. Markovian chains and the role of history in succession. *Trends in Ecology and Evolution* 5: 27–29. [8]

Fenchel, T. 1974. Intrinsic rate of natural increase: the relationship with body size. *Oecologia* 14: 317–326. [1]

Fisher, R. A. 1930. *The Genetical Theory of Natural Selection.* Clarendon Press, Oxford. [3]

Gadgil, M. and W. H. Bossert. 1970. Life historical consequences of natural selection. *The American Naturalist* 104: 1–24. [3]

Gallagher, E. D., G. B. Gardner and P. A. Jumars. 1990. Competition among the pioneers in a seasonal soft-bottom benthic succession: field experiments and analysis of the Gilpin–Ayala competition model. *Oecologia* 83: 427–442. [5]

Gause, G. F. 1934. *The Struggle for Existence*. Williams and Wilkins, Baltimore. [5]

Goodman, D. 1982. Optimal life histories, optimal notation, and the value of reproductive value. *The American Naturalist* 119: 803–823. [3]

Gotelli, N. J. 1991. Metapopulation models: the rescue effect, the propagule rain, and the core-satellite hypothesis. *The American Naturalist* 138: 768–776. [4]

Gotelli, N. J. and L. G. Abele. 1982. Statistical distributions of West Indian land-bird families. *Journal of Biogeography* 9: 421–435. [7]

Gotelli, N. J. and W. G. Kelley. 1993. A general model of metapopulation dynamics. *Oikos* 68: 36–44. [4]

Hanski, I. 1982. Dynamics of regional distribution: the core and satellite species hypothesis. *Oikos* 38: 210–221. [4]

Hanski, I. and M. Gilpin. 1991. Metapopulation dynamics: brief history and conceptual domain. *Biological Journal of the Linnean Society* 42: 3–16. [4]

Hardin, G. 1960. The competitive exclusion principle. *Science* 131: 1292–1297. [5]

Hardin, G. 1968. The tragedy of the commons. *Science* 162: 1243–1248. [2]

Harrison, S., D. D. Murphy and P. R. Ehrlich. 1988. Distribution of the bay checkerspot butterfly, *Euphydryas editha bayensis*: evidence for a metapopulation model. *The American Naturalist* 132: 360–382. [4]

Hilborn, R. and M. Mangel. 1997. *The Ecological Detective: Confronting Models with Data*. Princeton University Press, Princeton, N.J. [Preface]

Holling, C. S. 1959. The components of predation as revealed by a study of small mammal predation of the European pine sawfly. *Canadian Entomologist* 91: 293–320. [6]

Horn, H. S. 1975. Markovian processes of forest succession. *In* M. L. Cody and J. M. Diamond (eds.), *Ecology and Evolution of Communities*, pp. 196–213. Harvard University Press, Cambridge, Mass. [8]

Hudson, P. J., A. P. Dobson and D. Newborn. 1985. Cyclic and noncyclic populations of red grouse: a role for parasitism? *In* D. Rollinson and R. M. Anderson (eds.), *Ecology and Genetics of Host–Parasite Interactions*, pp. 77–89. Academic Press, London. [6]

Hudson, P. J., D. Newborn and A. P. Dobson. 1992. Regulation and stability of a free-living host–parasite system: *Trichostrongylus tenuis* in red grouse. I. Monitoring and parasite reduction experiments. *Journal of Animal Ecology* 61: 477–486. [6]

Huston, M. A. 1994. *Biological Diversity: The Coexistence of Species on Changing Landscapes*. Cambridge University Press, Cambridge. [8]

Hutchinson, G. E. 1967. *A Treatise on Limnology, Vol. II. Introduction to Lake Biology and Limnoplankton.* Wiley, New York. [5]

Iosifescu, M. and P. Tăutu. 1973. *Stochastic Processes and Applications in Biology and Medicine. Volume II. Models.* Springer-Verlag, Berlin. [1]

Keith, L. B. 1983. Role of food in hare population cycles. *Oikos* 40: 385–395. [6]

Kingsland, S. E. 1985. *Modeling Nature: Episodes in the History of Population Ecology.* University of Chicago Press, Chicago. [6]

Krebs, C. J. 1985. *Ecology: The Experimental Analysis of Distribution and Abundance,* 3rd Ed. Harper & Row, New York. [2,5]

Lack, D. 1967. *The Natural Regulation of Animal Numbers.* Clarendon Press, Oxford. [1]

Law, R. and R. D. Morton. 1993. Alternate permanent states of ecological communities. *Ecology* 74: 1347–1361. [8]

Lefkovitch, L. P. 1965. The study of population growth in organisms grouped by stages. *Biometrics* 21: 1–18. [3]

Leslie, P. H. 1945. On the use of matrices in certain population mathematics. *Biometrika* 35: 183–212. [3]

Levins, R. 1969. The effect of random variations of different types on population growth. *Proceedings of the National Academy of Sciences, USA* 62: 1061–1065. [2]

Levins, R. 1970. Extinction. *In* M. Gerstenhaber (ed.), *Some Mathematical Questions in Biology. Lecture Notes on Mathematics in the Life Sciences,* pp. 75–107. The American Mathematical Society, Providence, R.I. [4]

Lomolino, M. V. 1990. The target area hypothesis: the influence of island area on immigration rates of non-volant mammals. *Oikos* 57: 297–300. [7]

Luckinbill, L. S. 1979. Selection of the *r/K* continuum in experimental populations of protozoa. *The American Naturalist* 113: 427–437. [3]

MacArthur, R. H. 1972. *Geographical Ecology.* Harper & Row, New York. [5]

MacArthur, R. H. and E. O. Wilson. 1963. An equilibrium theory of insular zoogeography. *Evolution* 17: 373–387. [7]

MacArthur, R. H. and E. O. Wilson. 1967. *The Theory of Island Biogeography.* Princeton University Press, Princeton, N.J. [3,7]

May, R. M. 1973. *Stability and Complexity in Model Ecosystems.* Princeton University Press, Princeton, N.J. [2]

May, R. M. 1974a. Ecosystem patterns in randomly fluctuating environments. *Progress in Theoretical Biology* 3: 1–50. [1,2]

May, R. M. 1974b. Biological populations with non-overlapping generations: stable points, stable cycles, and chaos. *Science* 186: 645–647. [2]

May, R. M. 1976. Models for single populations. *In* R. M. May (ed.), *Theoretical Ecology: Principles and Applications,* pp. 4–25. W. B. Saunders, Philadelphia. [2]

McAuliffe, J. R. 1988. Markovian dynamics of simple and complex desert plant communities. *The American Naturalist* 131: 459–490. [8]

Mertz, D. B. 1970. Notes on methods used in life-history studies. *In* J. H. Connell, D. B. Mertz and W. W. Murdoch (eds.), *Readings in Ecology and Ecological Genetics*, pp. 4–17. Harper & Row, New York. [3]

Mitchell, W. A. and J. S. Brown. 1990. Density-dependent harvest rates by optimal foragers. *Oikos* 57: 180–190. [6]

Molofsky, J. 1994. Population dynamics and pattern formation in theoretical populations. *Ecology* 75: 30–39. [8]

Moran, P. A. P. 1949. The statistical analysis of the sunspot and lynx cycles. *Journal of Animal Ecology* 18: 115–116. [6]

Murphy, D. D. and P. R. Ehrlich. 1980. Two California bay checkerspot subspecies: one new, one on the verge of extinction. *Journal of the Lepidopteran Society* 34: 316–320. [4]

Murphy, G. I. 1968. Pattern in life history and the environment. *The American Naturalist* 102: 391–403. [3]

Noble, I. R. 1981. Predicting successional change. *In* H. A. Mooney (ed.), *Fire Regimes and Ecosystem Properties*. U.S. Department of Agriculture Forest Service General Technical Report WO-26. [8]

Pacala, S. W., C. D. Canham, J. Saponara, J. J.A. Silander, R. K. Kobe and E. Ribbens. 1996. Forest models defined by field measurements: estimation, error analysis, and dynamics. *Ecological Monographs* 66: 1–43. [8]

Pianka, E. R. 1970. On *r*- and *K*-selection. *The American Naturalist* 104: 592–597. [3]

Pielou, E. C. 1969. *An Introduction to Mathematical Ecology*. Wiley, New York. [1]

Pitt, D. E. 1993. Talks at U.N. combat threat to oceans' species from overfishing. *The New York Times*, 6/23/93. [2]

Polis, G. A., C. A. Myers and R. D. Holt. 1989. The ecology and evolution of intraguild predation: potential competitors that eat each other. *Annual Review of Ecology and Systematics* 20: 297–330. [5]

Ranta, E., J. Lindström, V. Kaitala, H. Kokko, H. Lindén and E. Helle. 1997. Solar activity and hare dynamics: a cross-continental comparison. *The American Naturalist* 149: 765–775. [6]

Ray, C., M. Gilpin and A. T. Smith. 1991. The effect of conspecific attraction on metapopulation dynamics. *Biological Journal of the Linnean Society* 42: 123–134. [4]

Real, L. 1977. The kinetics of functional response. *The American Naturalist* 111: 289–300. [6]

Reznick, D. N., M. J. Butler IV, F. H. Rodd and P. Ross. 1996. Life-history evolution in guppies (*Poecilia reticulata*). VI. Differential mortality as a mechanism for natural selection. *Evolution* 50: 1651–1660. [3]

Reznick, D. N., F. H. Shaw, F. H. Rodd and R. G. Shaw. 1997. Evaluation of the rate of evolution in natural populations of guppies (*Poecilia reticulata*). *Science* 275: 1934–1937. [3]

Roff, D. A. 1992. *The Evolution of Life Histories*. Chapman & Hall, New York. [3]

Rose, M. R. 1984. The evolution of animal senescence. *Canadian Journal of Zoology* 62: 1661–1667. [3]

Rosenzweig, M. L. 1971. Paradox of enrichment: destabilization of exploitation ecosystems in ecological time. *Science* 171: 385–387. [6]

Rosenzweig, M. L. and R. H. MacArthur. 1963. Graphical representation and stability conditions of predator–prey interactions. *The American Naturalist* 47: 209–223. [6]

Roughgarden, J. 1979. *Theory of Population Genetics and Evolutionary Ecology: An Introduction*. Macmillan, New York. [1,2,3]

Roughgarden, J. 1998. *Primer of Ecological Theory*. Prentice-Hall, Inc., Englewood Cliffs, N.J. [A]

Royama, T. 1971. A comparative study of models of predation and parasitism. *Researches in Population Ecology* (Supplement) 1: 1–91. [6]

Schoener, T. W. 1976. The species-area relation within archipelagoes: models and evidence from island birds. pp. 1–17 in *Proceedings of the 16th International Ornithological Congress*, Canberra, Australia. [7]

Schluter, D. 1994. Experimental evidence that competition promotes divergence in adaptive radiation. *Science* 256: 798–801. [3]

Simberloff, D. 1976. Species turnover and equilibrium island biogeography. *Science* 194: 472–478. [7]

Simberloff, D. S. and E. O. Wilson. 1969. Experimental zoogeography of islands: the colonization of empty islands. *Ecology* 50: 278–289. [7]

Sinclair, A. R. E. and J. M. Gosline. 1997. Solar activity and mammal cycles in the northern hemisphere. *The American Naturalist* 149: 776–784. [6]

Sinclair, A. R. E., J. M. Gosline, G. Holdsworth, C. J. Krebs, S. Boutin, J. N. M. Smith, R. Boonstra and M. Dale. 1993. Can the solar cycle and climate synchronize the snowshoe hare cycle in Canada? Evidence from tree rings and ice cores. *The American Naturalist* 141: 173–198. [6]

Slade, N. A. and D. F. Balph. 1974. Population ecology of Uinta ground squirrels. *Ecology* 55: 989–1003. [3]

Smith, C. H. 1983. Spatial trends in Canadian snowshoe hare, *Lepus americanus*: population cycles. *Canadian Field-Naturalist* 97: 151–160. [6]

Smith, J. N. M., C. J. Krebs, A. R. E. Sinclair and R. Boonstra. 1988. Population biology of snowshoe hares. II. Interaction with winter food plants. *Journal of Animal Ecology* 57: 269–286. [6]

Smith, J. N., P. Arcese and W. M. Hochachka. 1991. Social behaviour and population regulation in insular bird populations: implications for conservation. *In* C. M. Perrins, J.-D. Lebreton and G. J. M. Hirons (eds.), *Bird Population Studies: Relevance to Conservation and Management*, pp. 148–167. Oxford University Press, Oxford. [2]

Solomon, M. E. 1949. The natural control of animal populations. *Journal of Animal Ecology* 18: 1–35. [6]

Spear, R. W., M. B. Davis and L. C. K. Shane. 1994. Late quaternary history of low- and mid-elevation vegetation in the White Mountains of New Hampshire. *Ecological Monographs* 64: 85–109. [8]

Stearns, S. C. 1992. *The Evolution of Life Histories*. Oxford University Press, Oxford. [3]

Sutherland, J. P. 1974. Multiple stable points in natural communities. *The American Naturalist* 108: 859–873. [8]

Svane, I. 1984. Observations on the long-term population dynamics of the perennial ascidian, *Ascidia mentula* O. F. Muller, on the Swedish west coast. *Biological Bulletin* 167: 630–646. [2]

Tanner, J. E., T. P. Hughes, and J. H. Connell. 1996. The role of history in community dynamics: a modeling approach. *Ecology* 77: 108–117. [8]

Taylor, C. E. and C. Condra. 1980. *r* selection and *K* selection in *Drosophila pseudo-obscura*. *Evolution* 34: 1183–1193. [3]

Usher, M. B. 1979. Markovian approaches to ecological succession. *Journal of Animal Ecology* 48: 413–426. [8]

Werner, P. A. 1977. Colonization success of a "biennial" plant, teasel (*Dipsacus sylvestris* Huds.): experimental field studies in species cohabitation and replacement. *Ecology* 58: 840–849. [3]

Werner, P. A. and H. Caswell. 1977. Population growth rates and age vs. stage distribution models for teasel (*Dipsacus sylvestris* Huds.). *Ecology* 58: 1103–1111. [3]

Wiens, J. A. 1989. *The Ecology of Bird Communities, Vol. 1. Foundations and Patterns. In* R. S. K. Barnes, H. J. B. Birks, E. F. Connor, and R. T. Paine (eds.), Cambridge Studies in Ecology. Cambridge University Press, Cambridge. [8]

Williamson, M. 1981. *Island Populations*. Oxford University Press, Oxford. [7]

Wilson, E. O. and W. H. Bossert. 1971. *A Primer of Population Biology*. Sinauer Associates, Sunderland, Mass. [3]

Wilson, E. O. and D. S. Simberloff. 1969. Experimental zoogeography of islands: defaunation and monitoring techniques. *Ecology* 50: 267–295. [7]

Wolfram, S. 1984. Universality and complexity in cellular automata. *Physica* 10D: 1–35. [8]

Index

ABOUT THE AUTHOR

Nicholas J. Gotelli is Professor in the Department of Biology at the University of Vermont. He graduated with a B.A. from the University of California, Berkeley, and earned his Ph.D. at Florida State University. Dr. Gotelli currently serves on the Board of Editors of *Ecology*. His research interests include invertebrate community ecology, biogeography, island biology, plant demography, and the ecology of invasive species.

ABOUT THE BOOK

Editor: Andrew D. Sinauer
Project Editor: Carol J. Wigg
Production Manager: Christopher Small
Book and Cover Design: Jefferson Johnson
Book Production: Michele Ruschhaupt
Copy Editor: Roberta Lewis
Artwork: Precision Graphics
Book Manufacture: Courier Companies, Inc.